NEUROMETHODS

Series Editor
Wolfgang Walz
University of Saskatchewan
Saskatoon, SK, Canada

For further volumes:
http://www.springer.com/series/7657

Transmission Electron Microscopy Methods for Understanding the Brain

Edited by

Elisabeth J. Van Bockstaele

Department of Pharmacology and Physiology
Drexel University
Philadelphia, PA, USA

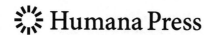 Humana Press

Editor
Elisabeth J. Van Bockstaele
Department of Pharmacology and Physiology
Drexel University
Philadelphia, PA, USA

ISSN 0893-2336   ISSN 1940-6045   (electronic)
Neuromethods
ISBN 978-1-4939-8103-8   ISBN 978-1-4939-3640-3   (eBook)
DOI 10.1007/978-1-4939-3640-3

Printed on acid-free paper

This Humana Press imprint is published by Springer Nature
The registered company is Springer Science+Business Media LLC New York

Preface to the Series

Experimental life sciences have two basic foundations: concepts and tools. The *Neuromethods* series focuses on the tools and techniques unique to the investigation of the nervous system and excitable cells. It will not, however, shortchange the concept side of things as care has been taken to integrate these tools within the context of the concepts and questions under investigation. In this way, the series is unique in that it not only collects protocols but also includes theoretical background information and critiques which led to the methods and their development. Thus it gives the reader a better understanding of the origin of the techniques and their potential future development. The *Neuromethods* publishing program strikes a balance between recent and exciting developments like those concerning new animal models of disease, imaging, in vivo methods, and more established techniques, including, for example, immunocytochemistry and electrophysiological technologies. New trainees in neurosciences still need a sound footing in these older methods in order to apply a critical approach to their results.

Under the guidance of its founders, Alan Boulton and Glen Baker, the *Neuromethods* series has been a success since its first volume published through Humana Press in 1985. The series continues to flourish through many changes over the years. It is now published under the umbrella of Springer Protocols. While methods involving brain research have changed a lot since the series started, the publishing environment and technology have changed even more radically. Neuromethods has the distinct layout and style of the Springer Protocols program, designed specifically for readability and ease of reference in a laboratory setting.

The careful application of methods is potentially the most important step in the process of scientific inquiry. In the past, new methodologies led the way in developing new disciplines in the biological and medical sciences. For example, Physiology emerged out of Anatomy in the nineteenth century by harnessing new methods based on the newly discovered phenomenon of electricity. Nowadays, the relationships between disciplines and methods are more complex. Methods are now widely shared between disciplines and research areas. New developments in electronic publishing make it possible for scientists that encounter new methods to quickly find sources of information electronically. The design of individual volumes and chapters in this series takes this new access technology into account. Springer Protocols makes it possible to download single protocols separately. In addition, Springer makes its print-on-demand technology available globally. A print copy can therefore be acquired quickly and for a competitive price anywhere in the world.

Saskatoon, Canada   *Wolfgang Walz*

Preface

Transmission electron microscopy (TEM) provides a powerful approach for advancing the understanding of the brain. Several books currently on the market focus on material science or approaches that emphasize techniques related primarily to scanning electron microscopy. The content of previously published books is often broad and applies to cell biologists at large who work with multiple organ systems. To date, there are virtually no books on TEM that specifically target neuroscientists. The goal of this NeuroMethods series is to provide the readership with detailed protocols and techniques focused on applied aspects of TEM for the study of neuroscience-based investigations.

The technique of TEM has evolved considerably over the past century. As summarized by *Zhang and colleagues*, although the design of the equipment itself has remained fundamentally similar, it is the advent of technical advances in specimen preparation including fixation, staining, embedding procedures, and availability of immunoreagents that has led to significant progress in the modern application of the technique. The authors provide a historical perspective on the evolution of the technology and review key developments that have resulted in sustained innovations.

As the specialized junction through which neurotransmitters are conveyed from one neuron to another, the synapse represents a key site for communication between brain cells. *Liu and Cheng* describe a detailed immuno-electron microscopy protocol for simultaneously labeling a presynaptically distributed neurotransmitter and a postsynaptically distributed one. Using a combination of pre-embedding immunoperoxidase labeling and post-embedding immunogold labeling, this dual immuno-EM labeling approach allows for reliable detection of differentially localized neurochemicals in closely related neural structures. The authors present key steps of the procedure where glutamate is labeled in the presynaptic terminal using immunogold labeling and CaM Kinase or ionotropic glutamate receptor subunits are labeled using immunoperoxidase detection in the postsynaptic structure. Taken with the resolution of the TEM that is capable of detecting even the smallest synaptic contacts, this method allows for clear detection of juxtaposition of different neurochemicals in brain circuits.

For the past several decades, TEM has been used to study neuronal architecture, from the first study to classify synapses in the cerebral cortex to detailed serial section analyses on large parts of identified neurons. However, it is the advent of serial section electron microscopy that has enabled nanoscale analysis of neuronal cell biology and anatomical connectivity. As described by *Bourne*, serial section EM provides the needed nanometer resolution to localize and measure synaptic connections, and identifies ultrastructural substrates of key cellular functions such as sites of local protein synthesis indicated by polyribosomes and local regulation of intracellular calcium and trafficking of membrane proteins by the network of smooth endoplasmic reticulum that extends throughout the neuron. Methods and procedures are described for obtaining, imaging, and analyzing serial sections of brain tissue. In addition, examples and descriptions are provided to help identify different types of spines and synapses and subcellular structures such as polyribosomes and smooth endoplasmic reticulum.

Experimental neuroanatomical tract tracing has elucidated the fine synaptic organization of neural circuitry in the brain. Although light microscopic analysis of tract tracers can provide important insight into selected neuronal circuits, its limitation lies in its inability to unequivocally demonstrate the presence of synapses between neurons in different regions. *Bajic* provides a detailed protocol for combining immunoperoxidase labeling of the antero-grade tracer biotinylated dextran amine with immunogold-silver enhancement detection of a neuroactive substance present in the structures postsynaptic to the tracer-labeled axon terminals. This technique is widely utilized to identify neuronal circuits as well as the neurochemical content of neurons in complex neural circuits.

The process of neuronal plasticity has been strongly supported by experiments involving TEM. As described by *Villalba and colleagues*, the use of computer-assisted 3D reconstructions of individual spines at the EM level has provided evidence that spines are highly plastic entities that are capable of complex structural remodeling in response to physiological or pathophysiological alterations. The authors describe protocols combining immuno-EM methods (to identify specific populations of presynaptic terminals or dendritic spines), serial ultrathin sectioning, and three-dimensional EM reconstruction to analyze ultrastructural and morphometric changes of individual dendritic spines in rhesus monkey models of brain disease. Using this approach, the authors have been able to quantify and compare various structural parameters of striatal spine morphology and specific glutamatergic synapses between normal monkeys and animal models of Parkinson's disease. Thus, using striatal projection neurons as a working model, the authors are able to conduct quantitative analyses to determine ultrastructural changes in the morphology, synaptic connectivity, and perisynaptic glial coverage of axo-spinous cortical or thalamic synapses between normal and MPTP-treated parkinsonian monkeys.

Understanding structural changes in the brain associated with neurodegenerative diseases has also been informed by ultrastructural studies involving invertebrate models. Using the fruit fly *Drosophila*, *Ando and colleagues* describe how this widely employed organism, known for its ease of genetic manipulation, has emerged as a powerful tool for studying human diseases. The authors describe the use of these transgenic models to recapitulate pathological phenotypes. Specifically, the authors provide a detailed approach for expressing Alzheimer's disease (AD) associated β-amyloid peptides or microtubule associated protein tau in the fly brain to gain an understanding of its pathological effects on the ultrastructure of the brain and within subcellular organelles. These analyses reveal that several critical pathologies observed in the brains of patients with AD are recapitulated in these fly models of the disease.

As mentioned above, the brain undergoes remarkable plasticity. However, brain injury causes a plethora of negative sequelae that can be revealed and better understood using TEM. *Simpson and Lin* describe the use of immunofluorescence staining in combination with ultrastructural analysis to reveal selective alteration of neurons and glial cells after early exposure to antidepressants and neurotoxins. The authors discuss how, with the advent of selective neurochemically, and/or receptor, specific antibodies as well as the refinement of fluorescent microscopy and digitized electron microscopic images, neuroscience researchers have an excellent opportunity to investigate maturation and plasticity of brain circuits as well as the developmental changes during early life. With such combined technical advancements, current knowledge on circuit function in the adult and dysfunction due to early exposure to environmental insults can contribute new knowledge.

Dual labeling immunohistochemistry employing visually distinct immunoperoxidase and immunogold markers has been an effective approach for elucidating complex receptor profiles at the synapse and for definitively establishing the localization of individual receptors and ligands to common cellular profiles. More recently, the combination of dual-immunogold-silver labeling of distinct antigens offers some unique benefits for resolving interactions between G-protein-coupled receptors, their interacting proteins, or downstream effectors. *Reyes and colleagues* describe an approach that provides superior subcellular localization of the antigen of interest while preserving optimal ultrastructural morphology. A detailed methodology is provided on the use of different-sized immunogold particles to analyze the association of the mu-opioid receptor with markers of early and late endosomes in neuronal intracellular compartments following systemic agonist exposure.

Determining the input/output characteristics of neurons with peripheral organs is fundamental to elucidating how the brain regulates physiological processes. *Havton and colleagues* have developed protocols to allow for EM detection of retrogradely labeled spinal cord neurons from the major pelvic ganglia or the rat external urethral sphincter muscle in combination with post-embedding immunogold labeling. Using Lowicryl HM20 for slow embedding of fixed brain tissues at low temperatures produces markedly improved antigen preservation. The protocol is versatile and can be combined with immunogold detection of neurotransmitters and membrane transporters so as to have broad applicability to ultrastructural studies of the central nervous system.

A number of experimental challenges previously existed with the detection of amino acids associated with synaptic transmission, including glutamate, L-aspartate, gamma-aminobutyric acid (GABA), and glycine. *Barnerssoi and May* present protocols for both retrograde and anterograde neuronal tracing that have been modified to facilitate electron microscopic examination of neuronal connectivity. In addition, it includes protocols for doing immunohistochemistry with antibodies to either glutaraldehyde-fixed GABA or glycine on ultrathin sections.

The visualization of these macromolecular complexes and assembly intermediates by TEM techniques can provide insights into their function in the cell as well as uncover their molecular mechanisms and life cycle. *Sander and Golas* describe how to prepare and analyze macromolecular assemblies by conventional negative staining EM, negative staining cryo-EM, and unstained cryo-EM. In particular, they focus on methods that allow imaging macromolecular assemblies that are near the size or concentration limits of the method or that are only transiently formed. Such challenging macromolecular assemblies can be subjected to the GraFix approach that cross-links the complexes under ultracentrifugation. Moreover, they also discuss EM imaging techniques and image processing approaches suited to compare macromolecular complexes captured at different conformational or compositional states by using difference mapping. Together, this set of single-particle EM methods provides an outline of how macromolecular complexes of the nervous system can be studied by EM techniques.

TEM has been extremely useful for visualizing microglial, oligodendrocytic, astrocytic, and neuronal subcellular compartments in the central nervous system at high spatial resolution. In some cases, neuronal damage is indirect. *Louboutin and colleagues* discuss methodological approaches directed at understanding HIV-1-Associated Neurocognitive Disorder (HAND), a neurodegenerative disease resulting in various clinical manifestations, characterized by neuroinflammation, oxidative stress, and related events. Neuronal damage occurs when microglial cells infected by HIV-1 increase the production of cytokines and release HIV-1 proteins, the most likely neurotoxins, among which are the envelope proteins gp120 and gp41, and the nonstructural proteins Nef, Rev, Vpr, and Tat. The authors present different methods used in the assessment of apoptosis and neuronal loss in different experimental, acute and chronic, models of HAND and consider how these techniques help to evaluate the effects of gene delivery of antioxidant enzymes in animal models of HAND.

In summary, each chapter in *NeuroMethods: Transmission Electron Microscopy for Understanding the Brain* is designed to provide detailed information on experimental protocols so that the novice can easily acquire the technique for their own objectives. The contributors are experts in TEM analysis and have all published their original research in peer-reviewed journals.

Philadelphia, Pennsylvania, USA   *Elisabeth J. Van Bockstaele*

Contents

Contributors

LOKESH AGRAWAL • *Department of Pathology, Anatomy and Cell Biology, Thomas Jefferson University, Philadelphia, PA, USA*

KANAE ANDO • *Department of Neuroscience, Thomas Jefferson University, Philadelphia, PA, USA; Laboratory of Molecular Neuroscience, Department of Biological Sciences, Tokyo Metropolitan University, Hachioji, Tokyo, Japan*

DUSICA BAJIC • *Department of Anesthesiology, Perioperative and Pain Medicine, Boston Children's Hospital, Boston, MA, USA; Department of Anaesthesia, Harvard Medical School, Boston, MA, USA*

MIRIAM BARNERSSOI • *Institute of Anatomy I, Ludwig-Maximilians University, Munich, Germany*

ELISABETH J. VAN BOCKSTAELE • *Department of Pharmacology and Physiology, College of Medicine, Drexel University, Philadelphia, PA, USA*

JENNIFER N. BOURNE • *Department of Physiology and Biophysics, University of Colorado Anschutz Medical Campus, Aurora, CO, USA*

HUIYI H. CHANG • *Department of Urology, University of Southern California, Los Angeles, CA, USA*

HWAI-JONG CHENG • *Center for Neuroscience, University of California at Davis, Davis, CA, USA*

MONIKA M. GOLAS • *Centre for Stochastic Geometry and Advanced Bioimaging, Aarhus University, Aarhus, Denmark; Department of Biomedicine, Aarhus University, Aarhus, Denmark*

LEIF A. HAVTON • *Department of Neurology, University of California, Los Angeles, Los Angeles, CA, USA*

STEPHEN HEARN • *St. Giles Foundation Advanced Microscopy Center, Cold Spring Harbor Laboratory, Cold Spring Harbor, NY, USA*

KOICHI M. IIJIMA • *Laboratory of Genetics and Pathobiology, Department of Alzheimer's Disease Research, National Center for Geriatrics and Gerontology, Obu, Aichi, Japan*

JANET L. KRAVETS • *Department of Pharmacology and Physiology, Drexel University, College of Medicine, Philadelphia, PA, USA*

RICK C.S. LIN • *Department of Neurobiology and Anatomical Sciences, University of Mississippi Medical Center, Jackson, MS, USA; Department of Psychiatry and Human Behavior, University of Mississippi Medical Center, Jackson, MS, USA; Department of Pediatrics, University of Mississippi Medical Center, Jackson, MS, USA*

XIAO-BO LIU • *Electron Microscopy Laboratory, Department of Cell Biology and Human Anatomy, School of Medicine, University of California at Davis, Davis, CA, USA*

JEAN-PIERRE LOUBOUTIN • *Department of Pathology, Anatomy and Cell Biology, Thomas Jefferson University, Philadelphia, PA, USA; Section of Anatomy, Department of Basic Medical Sciences, University of the West Indies, Kingston, Jamaica*

AKIKO MARUKO-OTAKE • *Department of Neuroscience, Thomas Jefferson University, Philadelphia, PA, USA*

PAUL J. MAY • *Department of Neurobiology and Anatomical Science, University of Mississippi Medical Center, Jackson, MS, USA; Department of Ophthalmology, University of Mississippi Medical Center, Jackson, MS, USA; Department of Neurology, University of Mississippi Medical Center, Jackson, MS, USA*

YI PANG • *Department of Pediatrics, University of Mississippi Medical Center, Jackson, MS, USA*

J.F. PARE´ • *Yerkes National Primate Research Center, Emory University, Atlanta, GA, USA; Udall Centers of Excellence for Parkinson's Disease Research, Atlanta, GA, USA*

BEVERLY A.S. REYES • *Department of Pharmacology and Physiology, College of Medicine, Drexel University, Philadelphia, PA, USA*

JENNIFER A. ROSS • *Department of Pharmacology and Physiology, College of Medicine, Drexel University, Philadelphia, PA, USA*

BJOERN SANDER • *Stereology and Electron Microscopy Laboratory, Department of Clinical Medicine, Aarhus University, Aarhus, Denmark; Centre for Stochastic Geometry and Advanced Bioimaging, Aarhus University, Aarhus, Denmark*

MICHIKO SEKIYA • *Laboratory of Genetics and Pathobiology, Department of Alzheimer's Disease Research, National Center for Geriatrics and Gerontology, Obu, Aichi, Japan*

KIMBERLY L. SIMPSON • *Department of Neurobiology and Anatomical Sciences, University of Mississippi Medical Center, Jackson, MS, USA; Department of Psychiatry and Human Behavior, University of Mississippi Medical Center, Jackson, MS, USA*

Y. SMITH • *Yerkes National Primate Research Center, Emory University, Atlanta, GA, USA; Udall Centers of Excellence for Parkinson's Disease Research, Atlanta, GA, USA; Department of Neurology, Emory University, Atlanta, GA, USA*

DAVID S. STRAYER • *Department of Pathology, Anatomy and Cell Biology, Thomas Jefferson University, Philadelphia, PA, USA*

EMIKO SUZUKI • *Laboratory of Gene Network, National Institute of Genetics, Mishima, Shizuoka, Japan; Department of Genetics, SOKENDAI, Mishima, Shizuoka, Japan*

VICTORIA TROVILLION • *Department of Pharmacology and Physiology, College of Medicine, Drexel University, Philadelphia, PA, USA*

R.M. VILLALBA • *Yerkes National Primate Research Center, Emory University, Atlanta, GA, USA; Udall Centers of Excellence for Parkinson's Disease Research, Atlanta, GA, USA*

XIN-MEI WEN • *Department of Pharmacology and Physiology, Drexel University, College of Medicine, Philadelphia, PA, USA*

LISA WULUND • *Department of Neurology, University of California, Los Angeles, Los Angeles, CA, USA*

JINGYI ZHANG • *Department of Pharmacology and Physiology, College of Medicine, Drexel University, Philadelphia, PA, USA*

Neuromethods (2016) 115: 1–20
DOI 10.1007/7657_2016_101
© Springer Science+Business Media New York 2016
Published online: 21 May 2016

Advances in Neuroscience Using Transmission Electron Microscopy: A Historical Perspective

Jingyi Zhang, Beverly A.S. Reyes, Jennifer A. Ross, Victoria Trovillion, and Elisabeth J. Van Bockstaele

Abstract

Unlike light and fluorescence microscopy techniques that may provide only limited resolution, transmission electron microscopy (TEM) allows enhanced subcellular precision by enabling high resolution of varied specimens. Although the first TEM was invented in 1931, the widespread use of TEM for biological studies did not start until the 1940's. From that time onward, TEM has revolutionized our knowledge and understanding of cellular processes. More importantly, the use of TEM has greatly advanced neuroscience research by defining the presence of synaptic specializations, the organization of synaptic vesicles, the identification of protein machinery in dendrites, and neural circuit organization. Combined with the use of autoradiography, immunocytochemistry, tract-tracing among others, the neurochemical signature of defined synaptic circuits have been characterized. Thus, with TEM's enormous investigative power, it will continue to serve as a major analytical tool in both physical and biological research. This Chapter describes seminal events utilizing TEM that have provided tremendous advances in field of neuroscience.

Keywords: Transmission electron microscopy (TEM), Neuroscience, Immunocytochemistry (ICC), Tract tracing, Cryo-electron microscopy (Cryo-EM), 3D reconstruction

1   Introduction

In 1931, Ernst Ruska and Max Knoll, two electrical engineers working at the Technological University of Berlin in Germany, developed the first transmission electron microscope (TEM). The TEM exploited the ability of electrons, which have a much shorter wavelength than photons, to provide greater resolution when compared with the light microscope [1, 2, 3]. Superior resolution capabilities enabled TEM users to analyze the fine details of any specimen of interest. Thus, with its enormous investigative power, the TEM continues to serve as a major analytical tool in both physical and biological scientific research.

Two key discoveries laid the foundation for the development of the TEM. Ernst Abbe described that the standard optical microscope resolution was limited by the wavelength of the light [4, 5]. Following Abbe's discovery, sustained efforts were made to overcome this limitation. German scientists August Köhler and Moritz von Rohr in the early part of the twentieth century developed a microscope that operated with ultraviolet light [6]. Although the microscope provided only twofold greater resolution when compared to traditional light microscopy and required expensive quartz lenses, its development provided the impetus to explore other illuminants in microscopy. Ultimately, this investigation led to the discovery of the electron beam as an illuminant capable of providing superior resolution when compared to traditional light and fluorescence microscopes [6, 7]. The second critical turning point in the development of the TEM occurred when German physicist Julius Plücker discovered that cathode rays (i.e., the electron beam) could be deflected by magnetic fields [8]. Utilizing this seminal observation, a research group led by Max Knoll built a device with primitive magnetic lenses that is considered to be the very first TEM in the world. Soon after, aided by the De Broglie hypothesis, which revealed the wave nature of electrons, the same group built the first applied TEM with magnifications exceeding those of the light microscope in 1933 [1, 9].

TEM essentially consists of an electron gun, a series of apertures, and a set of magnetic lenses (Fig. 1). High-voltage electrons are emitted by the electron gun and form a beam controlled by the magnetic lenses. The electron beam is first adjusted by apertures and then passed through the specimen. Subsequently, the electron beam is focused onto a fluorescent screen or captured as an image, originally on film and now by a light sensitive sensor, such as a charge-coupled device camera. Hence, TEM, as the name implies, forms images by capturing an electron beam transmitted through the sample of interest [2, 3].

A significant challenge in refining the TEM and its technology was the need for samples to be prepared sufficiently thin to allow for the electrons to penetrate the specimen, yet be sufficiently strong so as to resist the vacuum and radiation damage produced by the electron beam. Thus, the evolution of TEM technology was accompanied by parallel advancements in sample preparation methods. At the beginning, chemical fixation followed by metal or negative staining was the standard methodology for TEM sample preparation for the analysis of biological specimens. The metal shadowing technique was established in the very early days of biological TEM study and was followed by the negative staining approach [10, 11]. During the early 1980s, the team of Jacques Dubochet developed the plunge-freezing cryo-technique, which enabled researchers to observe unstained biological specimens in their natural aqueous state using TEM [12, 13].

Different image capturing techniques are used by different types of electron microscopes. The TEM forms images by capturing an electron beam transmitted through the samples. The scanning

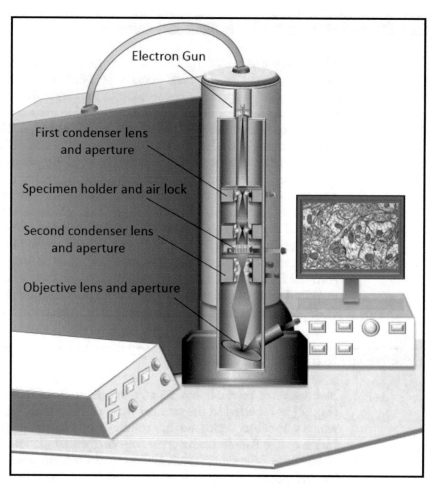

Fig. 1 Components and organization of a transmission electron microscope (TEM)

transmission electron microscope (STEM) is a type of TEM, but it is distinguished from conventional TEM by scanning the sample using a focused electron beam. Thus, it simultaneously collects the information from electron beams transmitted through the sample and various signals released from the sample surface. Using STEM, high-contrast imaging of biological samples can be achieved without staining. Unlike TEM, scanning electron microscopy (SEM) produces images by collecting signals reflected from the sample surface. Although SEM has a lower resolution compared to TEM, it can provide an acceptable representation of the three-dimensional shape of a sample. In addition, the environmental SEM, which can be utilized for wet sample and operated under lower vacuum conditions, enables researchers to analyze samples vulnerable to high vacuum conditions at the EM level [3]. While different types of electron microscopes offer varying advantages, this chapter focuses primarily on the application of TEM to brain tissue analysis.

2   The Early Stage of TEM Application in Biological Samples

The widespread use of the TEM in biological studies did not start until the 1940s. In the early periods of TEM use for biological sample analysis, applications evolved from the simple observation of rudimentary images of viruses to the examination of a single layer of cells that had been chemically fixed and dried [14, 15]. In 1949, the thin sectioning and plastic embedding of samples made it possible to explore the cell's internal structure [16, 17]. This progress led to a surge in publications regarding the fine structure of cellular organelles [18–20] as well as numerous reports showing details of cell membrane systems [21–23]. These initial ultrastructural studies played a pivotal role in establishing the foundation for our basic understanding of cellular organization and morphology. The importance of TEM during this time period is reflected by the study of Huxley and colleagues, whose TEM analysis revealed the myofibril array of skeletal muscle cells and provided the first direct evidence for the sliding filament theory of muscle contraction [24].

3   TEM Contributions to Neuroscience Research

Neuroscience research has greatly benefited from TEM studies (Fig. 2) as reflected in the seminal work of Palay and Palade [25], which was subsequently further solidified by the significant publication of *The Fine Structure of the Nervous System* in 1970 [26]. The observation of synaptic specializations provided unequivocal proof for the neuron doctrine that the nervous system is made up of discrete individual cells rather than an anatomical continuum [25].

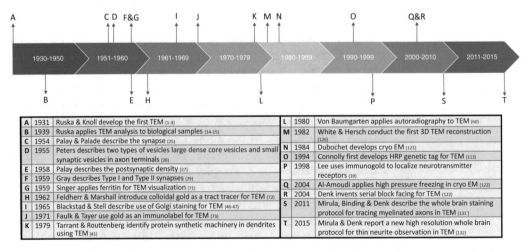

A	1931	Ruska & Knoll develop the first TEM (1-3)
B	1939	Ruska applies TEM analysis to biological samples (14-15)
C	1954	Palay & Palade describe the synapse (25)
D	1955	Peters describes two types of vesicles large dense core vesicles and small synaptic vesicles in axon terminals (26)
E	1958	Palay describes the postsynaptic density (27)
F	1959	Gray describes Type I and Type II synapses (29)
G	1959	Singer applies ferritin for TEM visualization (71)
H	1962	Feldherr & Marshall introduce colloidal gold as a tract tracer for TEM (72)
I	1965	Blackstad & Stell describe use of Golgi staining for TEM (46-47)
J	1971	Faulk & Tayer use gold as an immunolabel for TEM (73)
K	1979	Tarrant & Routtenberg identify protein synthetic machinery in dendrites using TEM (41)

L	1980	Von Baumgarten applies autoradiography to TEM (60)
M	1982	White & Hersch conduct the first 3D TEM reconstruction (126)
N	1984	Dubochet develops cryo EM (121)
O	1994	Connolly first develops HRP genetic tag for TEM (113)
P	1998	Lee uses immunogold to localize neurotransmitter receptors (39)
Q	2004	Al-Amoudi applies high pressure freezing in cryo EM (122)
R	2004	Denk invents serial block facing for TEM (122)
S	2011	Mirula, Binding & Denk describe the whole brain staining protocol for tracing myelinated axons in TEM (131)
T	2015	Mirula & Denk report a new high resolution whole brain protocol for thin neurite observation in TEM (132)

Fig. 2 Timeline of key milestones in neuroscience research arising from the use of the transmission electron microscope (TEM)

TEM studies also confirmed the existence of dendritic spines which had only been previously observed in Golgi-impregnated neurons [27]. High-resolution TEM analysis of axonal terminals offered an improved understanding of neurotransmitter storage and release. Two types of vesicles were described in axon terminals, small synaptic vesicles (SSV) later shown to accumulate fast-acting transmitters, glutamate and gamma amino butyric acid (GABA), and large dense core vesicle (LDCV), later shown to be the primary site for neuropeptide storage [26]. Quantitative immunogold studies further revealed that mammalian central nervous system (CNS) axon terminals, which form asymmetric and symmetric junctions correspond to excitatory and inhibitory synapses, respectively [26–28]. This general rule of asymmetrical synapses being excitatory in nature, while symmetrical synapses are inhibitory does not appear to hold for all neurotransmitters, especially monoamines and acetylcholine [29, 30]. For instance, although acetylcholine has mainly excitatory effects on the basolateral amygdala (BLa), almost all cholinergic synapses in the BLa are of the symmetrical type [30–32]. Monoamine transmitters can be stored in either SSV or LDCV or both [33]. Unlike SSV, which are primarily released from classical synaptic junctions, TEM studies supplied a wealth of evidence for putative extrasynaptic release of the contents of LDCVs [33, 34].

While one of the key contributions of TEM to neuroscience research was to demonstrate the existence of synapses, TEM also provided critical evidence for non-synaptic transmission or volume transmission [35]. Studies of receptor distribution established the existence of neuropeptide receptors at a distance from synaptic junctions, raising the possibility of extrasynaptic binding and activation [28], and strongly supporting the theory of volume transmission. For example, mu opioid receptor immunoreactivity has been found at extrasynaptic sites on noradrenergic dendrites in the rat locus coeruleus (LC) [36]. Likewise, monoamine neurotransmitters show the same tendency for volume transmission, as evidenced by low synaptic incidence of their terminals and extrasynaptic localization of their receptors [29, 37, 38].

Another important contribution to neuroscience made by TEM studies was to provide the morphological basis for the concept of synaptic plasticity. Identifying protein synthesis machinery in dendrites provided evidence of dendritic protein synthesis, indicating that dendrites are not only involved in signal transduction, but also participate in the long-term modulation of synaptic efficacy [39–42].

4 Golgi-EM, Molecular Labeling, and TEM

In 1873, an Italian physician, Camillo Golgi, established a silver staining method called "Golgi impregnation" for neuroanatomical analysis, which allowed, for the first time, observation of all parts of an individual neuron including the cell body, axon, dendrites, and dendritic spines [43]. Since then, this method has been continually used for whole neuron visualization, dendritic arborization analysis and spine quantification using light microscopy [44]. Although TEM had been introduced into biological research as early as the 1940s, it was not until 1965 that the Golgi precipitate was found to be an electron dense substance and became utilized as an electron microscopic marker [45, 46]. Combining Golgi impregnation with TEM promoted the advancement of the study of neuronal microcircuits. One of the major breakthroughs resulting from the use of Golgi staining in TEM was the identification of the postsynaptic neuronal profiles of severed axonal afferent projections [45]. The Golgi method was continually refined to better serve TEM studies. Because the ultrastructure in Golgi stained samples was often masked by the Golgi precipitate and hard to recognize, Blackstad established a "de-impregnation" protocol to partially remove Golgi precipitate by ultraviolet light illumination and thereby reveal the underlying fine structure of cells [47]. Soon after, Fairén invented the gold toning method for de-impregnating Golgi-stained sections based on Blackstad's work [44, 48]. Fairén and Blackstad then worked together to optimize the gold toning method by illuminating sections with strong white light before gold toning [44].

In 1983, Freund and Somogyi contributed further advances to the application of Golgi staining to TEM by applying silver impregnation to tissue sections [49, 50]. Through technical refinements to the histological procedure, Golgi impregnation can now be combined with a variety of other techniques, such as tract tracing, immunocytochemistry, and intracellular dyes [43, 44].

Today, intracellular injection of tracers or gene manipulation provides a new way of visualizing neurons and their processes in their entirety. Although Golgi-EM is less popular than when it was first conceived, Golgi staining, as well as its numerous variants, remains a widely used tool for neuroanatomical observations at light microscope level [51–57].

5   Immunocytochemistry and TEM

The capricious nature of Golgi staining limited its application as a reliable and consistent EM marker. This drove scientists to seek more specific detection methods. The first milestone occurred at the end of 1950s when Yalow and Berson introduced the radio immuno-assay [58]. The application of autoradiography in TEM provided the fundamental observations of the distribution of certain receptors, enzymes, or nucleic acids in brain sections prior to the emergence of commercially available antibodies [59, 60]. However, long exposure times, restricted sensitivity, complexity of handling, and disposal of isotopes hampered the broad application of the autoradiography procedure. With the emergence of the antibody production industry, immunocytochemistry (ICC) began to dominate the immunoassay field from the mid-1990s onward.

Cell-specific ICC provided an easy way to recognize certain cell types in the brain. For example, tyrosine hydroxylase could be used as a marker for catecholaminergic neurons. Similarly, the neuro-chemical signature of axon terminals could be defined by immuno-cytochemical approaches. In conjunction with TEM, ICC could reveal the subcellular localization of any molecule or structure of interest, as long as an antibody against the antigen was available.

Experiments involving ICC utilize specific antibodies to probe an antigen of interest in brain tissue. This antigen-antibody binding is then visualized using a detectable ligand conjugated to the anti-body. The ligand can be a detectable marker such as a fluorochrome, an electron-dense particle, or an enzyme that catalyzes a chemical reaction converting a substrate into an observable product. Peroxi-dase, especially horseradish peroxidase (HRP) is the most widely used reporter enzyme for TEM studies [61]. Avidin-biotin complex was added into the procedure for amplifying the signal in order to visualize antigens in low abundance [62]. A variety of chromogenic substrates for HRP have been developed and these chromogens are able to produce precipitates of different colors [62]. Among them, diaminobenzidine (DAB) is the most sensitive and its precipitate exhibits pronounced osmiophilicity, rendering the chromogen DAB as a popular choice in standard TEM protocols [63].

There are also some commercially available products with pro-prietary licenses that produce different colored reaction products compared to the common chromogenic substrates like DAB, for example, Vector VIP which exhibits an intense purple color [62]. DAB and Vector VIP exhibit different characteristics under TEM. DAB is diffuse in appearance within the labeled profiles, while the Vector VIP reaction product is particulate, facilitating differentia-tion of one from the other. Therefore, DAB and Vector VIP are able to be used together for dual labeling TEM studies, such as when examining the norepinephrine innervation of calmodulin-dependent protein kinase-labeled pyramidal cells in the rat BLa [29].

Although the fluorescent method has a higher sensitivity com-pared to chromogenic protocols, the fluorescent dyes exhibit poor osmiophilicity and require a multiple-step process to enhance their osmiophilicity for TEM investigation [64]. Maranto found that Lucifer Yellow (LY) in the presence of DAB converts into a dark product, which is visible in both light and electron microscope [65]. Later on, this DAB photoconversion was found to be a common phenomenon and could be apply to different fluorescent dyes [66–68]. However, this extra process often reduces tissue integrity at the ultrastructural level, and therefore limits the application of fluorescent labeling in TEM research [62]. Today, this photo-oxidation has been utilized for developing genetic tags and thus has become a powerful tool to provide precise spatial information on the location of fluorescence-labeled molecules (*see* Sect. 8).

Aside from catalytic reactions, antigen visualization can also be achieved by directly conjugating electron dense labels to the antibody, such as iron-rich protein ferritin or colloidal gold. This was first implemented in 1959 by Singer, who successfully conjugated antibodies to ferritin [69]. Due to its restricted penetration and nonspecific adsorption, ferritin is seldom used today. In contrast, because of its high spatial resolution, colloidal gold has become a very popular TEM marker. Colloidal gold was introduced into TEM research as a tracer in the early 1960s [70], but it was not applied as a label for antisera until 1971 [71]. With a highly visible electron-dense product in the form of punctate particles, colloidal gold can be easily detected using TEM and readily quantified. Moreover, gold particles can be prepared in different sizes ranging from 1 to 150 nm and therefore can be utilized for double labeling or even triple labeling, if combined with DAB.

Immunogold combined with TEM is a powerful tool to localize the distribution of neurotransmitter receptors and their associated proteins. Different distribution patterns of different subtypes of receptors imply their different functions. For instance, in rat locus coeruleus, alpha 2C adrenergic receptors are predominantly localized postsynaptically, while alpha 2A adrenergic receptors are localized to both presynaptic and postsynaptic structures [38, 72]. Since immunogold can provide adequate spatial resolution to differentiate membrane versus intracellular location, it became an ideal approach for G protein-coupled receptor trafficking studies. For example, a series of TEM studies has revealed a highly dynamic distribution of corticotrophin-releasing factor receptors [73].

6   Tract Tracing and TEM

Immunolabeling combined with TEM offers a wealth of information regarding the localization of specific molecules, neurochemical features of pre and post synaptic profiles and morphological changes in response to stress, injury and disease. Thus, it provides an intricate anatomical view of a brain area at the subcellular level. However, in order to fully understand the synaptic organization of the brain, experimental approaches directed at examining connectivity between brain regions were required. To this end, axonal tracers became valuable tools for investigating neuronal circuitry.

In tract-tracing experiments, locally injected tracers are absorbed endocytically, either by the soma or dendrites of neurons and then transported to their axon terminals (anterograde tracing), or they are taken up by axon terminals and then transported back to the soma (retrograde tracing). Their subcellular localization in the target area can be visualized using TEM and detection methods discussed above. Retrograde transport of HRP was introduced in 1970 [74] and its extension to the TEM level was described in 1984 [75]. Subsequently, Mesulam and colleagues in 1978 [76] applied the tracer wheat germ agglutinin (WGA)-HRP (capable of being transported both anterogradely and retrogradely) for TEM [77, 78]. Fluorogold is another valuable retrograde tracer that has been used for both light and TEM analysis [79–84]. Phaseolus vulgaris leucoagglutinin (PHA-L), a kidney bean lectin, was first described in 1978 [85] and subsequently used as a neuroanatomical tool for tracing connections in the CNS based on its predominantly anterograde axonal transport properties when delivered iontophoretically [86]. Biotinylated dextran amine was also developed as an efficient anterograde tracer in tracing studies [87–90].

Recently, genetic constructs have also been used as effective tract tracers. The most frequently used genetic tracers are recombinant viruses, which travel across synapses and therefore are able to define the simultaneous connectivity of multiple synapses [91–93]. Numerous pathways between brain regions have been identified through tract tracing methods [94]. Tract tracing investigations combined with immunocytochemistry have revealed that many fiber tracts in these neural pathways are anatomically and neurochemically heterogeneous [95, 96].

The source of presynaptic terminals to a given brain region can be localized via tract tracing. For example, research utilizing double labeling of GABA and retrograde tracer fluorogold revealed that GABA-containing neurons in the ventral tegmental area project to the nuclear accumbens in rat brain [97]. In addition to identifying pre and post-synaptic structures, one of the most advantages of using TEM has been the ability to unequivocally identify synaptic specializations between neurons. With the help of immunolabeling and tract tracing, the synaptology of defined brain circuits can be clearly established using TEM analysis. For instance, triple labeling ultrastructural studies showed that the prefrontal cortex selectively projects to GABA-containing, nucleus accumbens projecting neurons and to dopamine-containing, prefrontal cortex-projecting neurons in the ventral tegmental area [98].

7   Intracellular Recording and TEM

Electrophysiological investigations combined with TEM studies indicate that neurons are both physiologically and morphologically heterogeneous. Combining the information provided by these two distinct approaches has enabled a better understanding of the dynamic properties of brain circuits [99]. Successful strategies utilized to ascribe function to selected neuronal populations involved parallel studies of intracellular recording and TEM analysis [100–102]. Several intracellular dyes are now available for this strategy, such as HRP, biocytin, fluorescence dye Lucifer yellow, and ethidium bromide [103]. Following the electrophysiological recording of a neuron, an intracellular dye is injected into the recorded neurons, enabling the analysis of how complex synaptic responses relate to the neuron's morphology and its neurochemical signature [104–106]. By combining the anatomical and electro-physiological approaches, detailed evidence of the functional cross talk between different neurotransmitter systems can be revealed [106]. Despite the power of the combined approach, it has had limited use because the in vivo intracellular technique is quite challenging [104, 107, 108]. Today, most intracellular staining is done in vitro, using slice preparations. However, as preserving the integrity of the subcellular milieu in a slice preparation is challeng-ing, poor ultrastructural preservation has limited the usefulness of this intracellular labeling approach for TEM.

8   Application of Genetic Tags in the TEM Studies

To obtain optimal ultrastructural preservation and penetration of antibody in tissue sections for analysis, a balance of fixation and immunocytochemistry conditions must be achieved. The ultimate goal is to allow for the best penetration of antibodies, while main-taining a high quality of the ultrastructure, which is the paramount challenge for electron microscopists. Many procedures used during light microscopy sample preparation will degrade the quality of the tissue; thus fixation is a critical step during sample preparation. Acrolein and glutaraldehyde have been used for TEM-level fixation because of their outstanding ability for stabilizing cellular compo-nents and proteins. However, fixations will degrade the antigenicity and therefore interfere with optimal immunolabeling [109, 110]. To circumvent this dilemma, molecular approaches have been devel-oped to produce genetically engineered tags. Using this method, the target protein is specifically marked by genetic fusion and can be detected using the TEM. A horseradish peroxidase tag was first developed to investigate the morphological basis of exocytotic traf-ficking, however the high-calcium condition required limited its application [111]. Recently, a fluorescent protein tag "MiniSOG" was engineered. Illumination of this tag will generate oxygen, which can locally polymerize DAB into a reaction product resolvable by TEM [112]. Another tag recently developed genetic tag is a peroxi-dase called "APEX," which does not need light to polymerize DAB [113]. In addition, some fluorescent markers can be photo-oxidized to generate electron contrast. In this way, fluorescence image taken in living tissues could be correlated with the micrograph obtained by TEM from thin sections of the same area [114].

9   Unstained Cryo-Electron Microscopy (Cryo-EM)

As mentioned above, a primary disadvantage of using TEM is the requirement for the specimen to be sufficiently thin in order to be transparent to electron beam. Pursuant to multiple processing steps, cellular structures can be altered during the preparation process. In addition, the fixation and staining can introduce artifacts. In the early 1970s, a considerable amount of effort was generated to minimize such occurrences. At that time, different types of chambers for holding the sample were developed to observe wet unstained samples under TEM. However, none of them were ultimately useful because of poor resolution and sample damage [115].

A major breakthrough occurred in the early 1980s when two groups found that small water droplets could be vitrified by rapid cooling [116–118]. Based on this phenomenon, Jacques Dubochet and colleagues in 1984 developed the thin film vitrified technique and cryo-EM became a reliable tool for structural molecular biology. However, this method can only be applied to samples thin enough to fit in the thin liquid film, such as virus suspension, purified organelles and thin regions of whole cells [119]. Cryo-EM of vitrified sections was not available until 2004 because of the inevitable ice crystal formation in thicker samples [120]. The advent of high-pressure freezers, provided an effective way of vitrifying samples of greater volumes by applying liquid nitrogen with very high pressure to prevent ice crystal damage [121].

Compared with traditional TEM using chemical fixation, cryo-EM can better preserve tissues and cells in their natural states and therefore provide more reliable information about their ultrastructure [122]. Studies comparing aldehyde perfusion with cryo-fixation found that cryo-fixation preserves a physiological extracellular space, reveals larger numbers of synaptic vesicles, less intimate glia coverage, and smaller glial volume compared to chemical fixation [122]. An elegant protocol for the cryo-EM technique with detailed information has been published recently [123].

10   Three-Dimensional Reconstruction and EM

The advancement of information technology has largely expanded the capability of TEM techniques by providing targeted software capable of rapidly and accurately processing large amounts of data, leading to the development of three-dimensional (3D) EM reconstruction. 3D reconstruction of serial TEM sections was achieved first through a painstaking manual approach, which provided valuable information about how neurons are interconnected [124]. However, this approach was not practical for larger samples. The invention of automated serial sectioning overcame this issue, allowing for the collection of thousands of thin sections in series for TEM imaging [125]. An alternative technique to acquire tomographic images via TEM involved the capture of images in different orientations, whereby the sections are rotated around an axis perpendicular to the electron beam. The electron micrographs acquired from different angles are adjusted for positioning and focusing error before reconstruction [126]. The development of serial block-face scanning electron microscopy provided a much easier way to examine serial sections and complete reconstruction [127]. A microtome is built into its recording chamber, allowing for image acquisition after each cut [128]. The merit of this new technique is to directly image the large tissue sample itself without dividing it into smaller pieces, thereby avoiding damage caused by cutting and making it easy to follow the neural processes during reconstruction. A whole-brain staining method was developed to meet the needs of this serial block-face scanning electron microscope. Mirula et al. described a whole mouse brain staining and embedding procedure in 2012, which could be used to trace myelinated axons with low error rates [129]. Recently, Mirula and Denk reported a new high-resolution whole-brain procedure, which provides a well preserved preparation to identify the synapses and thin neurite such as spine necks [130]. All of the reconstruction methods listed above need specially designed software to computationally merge the images into 3D reconstructions. Another promising direction for 3D-EM is correlative microscopy, in which light microscopy or confocal microscopic images are used as a guide for analyzing reconstruction data.

Similar to TEM, Cryo-EM can also be adapted to 3D-EM. In 3D cryo-EM, cryo-fixed samples are directly visualized using the TEM and generate 2D images corresponding to a projection of the structure. The 3D images are then constructed by combining these 2D images in different perspectives [131]. Cryo-electron tomography produces 3D imaging of molecular components in their native status, which provides valuable information about the native architecture of the molecular machinery of the neuron [132]. Currently, cryo-EM has only been successfully conducted in cultured neurons [133, 134].

The emergence of 3D-EM reconstruction has allowed for the in-depth spatial and morphological analysis of the brain microenvironment in pathological disease states that are complex and varied in nature, e.g., in Alzheimer's disease (AD). The pathophysiology of AD is thought to arise, in part, from protein misfolding and hindered clearance of aggregated proteins. Illustrative of this point are the hallmark features of AD, including senile plaques, which are composed of amyloid beta (Aβ) peptide aggregates, and of neurofibrillary tangles (NFT), composed of hyperphosphorylated tau aggregates [135] (Fig. 3). Synaptic loss is a prominent feature of

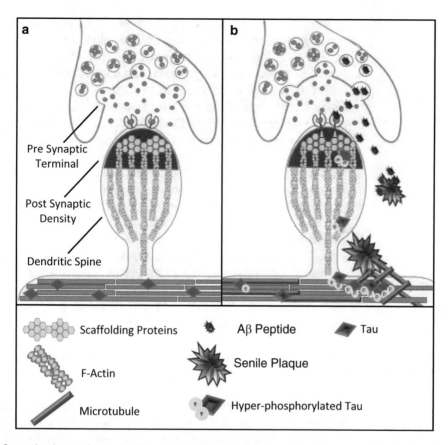

a

Pre Synaptic Terminal

Post Synaptic Density

Dendritic Spine

b

Scaffolding Proteins

F-Actin

Microtubule

Aβ Peptide

Senile Plaque

Hyper-phosphorylated Tau

Tau

Fig. 3 Synaptic dysregulation in Alzheimer's disease. Panel **a** shows an example of normal synaptic organizations. In Panel **b**, Amyloid beta (Aβ) peptides and hyper-phosphorylated tau aggregate to form senile plaques and neurofibrillary tangles, respectively. While the mechanisms of such interactions are still under investigation, it is likely that alterations in synaptic structure and function induced by senile plaques and neurofibrillary tangles ultimately lead to synapse loss

AD, and it is currently the best correlate of dementia severity [136]. Thus, the ability of 3D-EM reconstruction to estimate synaptic density, and to depict, in great detail, the architecture of the synapse distinguishes it as an invaluable tool in understanding the disease state and its progression [137, 138]. AD is also known to have an inflammatory component in which microglia and astrocytes surrounding affected neurons and Aβ plaques are activated, engaging the brain inflammatory response. These events are thought to exacerbate cognitive deficits in AD patients. Thus, a critical component in advancing our understanding of this disease state is to better understand the organization and structure of the brain milieu under normal and pathological conditions.

In this regard, 3D-EM reconstruction has shed light on mechanisms of interaction between these primary components of AD and the cellular and subcellular organization of neurons under conditions of neurodegeneration [139]. Nuntagij and colleagues utilized 3D reconstruction EM in both 3xTg mice and naturally aged dogs to extensively describe the close spatial relationship between Aβ deposits and the neutropil, revealing entangled and branched plaques that engulf soma and apical dendrites [140]. The ability of 3D-EM reconstruction to portray the complex disruptive forces of Aβ plaques and NFT are evident in the work of Fiala and colleagues, which describes the disruption of subcellular organization and function of mitochondria in the aged monkey brain [141]. The study results and analysis describe swollen, dystrophic neurites within late stage plaques. These neurites display pouches that are impaled with microtubules to form loops that trap mitochondria and other organelles, rendering them non-functional. Thus, 3D-EM analysis has been instrumental in defining the substrates of Aβ and NFT interaction and the abnormal organization of subcellular structures. Importantly, 3D-EM reconstruction has played an integral role in establishing that intraneuronal Aβ aggregation may disrupt intracellular transport, leading to the aberrant accumulation and dysfunction of mitochondria, potentially resulting in autophagic degeneration [139].

11   Future Perspectives

Imaging and data analysis techniques have undergone significant advancement since the TEM was invented 80 years ago. In the future, development of detectors with higher sensitivity and better signal/noise ratio would further reduce the beam intensity without sacrificing the resolution. The duration of time required to acquire images from large samples is an obstacle that needs to be resolved. The emergence of the quick imaging camera may eventually solve this problem. Another challenge that the current automatic reconstruction technique is facing a high error rate, which now can only be compensated for by time-consuming manual proofreading. Perhaps the greatest hurdle facing modern 3D-EM reconstruction is the time required for data analysis and the lack of appropriately specialized programs to complete automated analysis.

While there is much progress to be made in the technical aspects of TEM and 3D-EM reconstruction, the potential application of this technology has tremendous utility for our understanding of complex mental illnesses such as AD and other neuropathological conditions. The capacity to investigate synaptic density, functional characteristics of the presynaptic terminal, cleft, and post synaptic density, and the ability to provide crucial morphological information uniquely positions TEM, and more specifically 3D-EM reconstruction, for the intensive analysis of brain states that define a plethora of neurological disorders with various etiologies. Thus, innovation in this technology in parallel with computer science programming advances will facilitate new discoveries in the field of neuroscience.

Acknowledgements

This work was supported by PHS grants DA09082 and DA020129. We are grateful for the valuable information provided by Zeiss and FEI companies.

References

1. Nobelprize.org (2014) The Nobel Prize in Physics 1986—Perspectives. Web
2. Knott G, Genoud C (2013) Is EM dead? J Cell Sci 126(Pt 20):4545–4552
3. Jensen EC (2012) Types of imaging, Part 1: Electron microscopy. Anat Rec (Hoboken) 295(5):716–721
4. Abbe E (1873) Beiträge zur Theorie des Mikroskops und der mikroskopischen Wahrnehmung—I. Die Construction von Mikroskopen auf Grund der Theorie. Arch Mikrosk Anat 9(1):6
5. Abbe E (1876) A contribution to the theory of the microscope and the nature of microscopic vision. In: Proceedings of the Bristol Naturalists' Society. Williams & Northgate, London, UK
6. Ultraviolet microscope (2015) In Encyclopædia britannica. http://www.britannica.com/technology/microscope/The-theory-of-image-formation
7. Ruska E (1980) The early development of electron lenses and electron microscopy. S. Hirzel, Stuttgart, Translation by T Mulvey. ISBN 3-7776-0364-3
8. Plücker J (1858) Über die Einwirkung des Magneten auf die elektischen Entladungen in verdünnten Gasen [On the effect of a magnet on the electric discharge in raarified gases]. Poggendorffs Annalen der Physik und Chemie 103:88–106
9. Broglie LD (1928) La nouvelle dynamique des quanta. In Électrons et Photons: Rapports et Discussions du Cinquième Conseil de Physique, Solvay
10. Bradley DE (1958) Simultaneous evaporation of platinum and carbon for possible use in high-resolution shadow-casting for the electron microscope. Nature 181 (4613):875–877
11. Brenner S, Horne RW (1959) A negative staining method for high resolution electron microscopy of viruses. Biochim Biophys Acta 34:103–110
12. Adrian M et al (1984) Cryo-electron microscopy of viruses. Nature 308(5954):32–36
13. McDowall AW et al (1984) Cryo-electron microscopy of vitrified insect flight muscle. J Mol Biol 178(1):105–111
14. Baumeister W (2005) A voyage to the inner space of cells. Protein Sci 14(1):257–269
15. Kruger DH, Schneck P, Gelderblom HR (2000) Helmut Ruska and the visualisation of viruses. Lancet 355(9216):1713–1717
16. Newman SB, Borysko E, Swerdlow M (1949) New sectioning techniques for light and electron microscopy. Science 110(2846):66–68
17. Porter KR, Blum J (1953) A study in microtomy for electron microscopy. Anat Rec 117 (4):685–710
18. Palade GE (1952) The fine structure of mitochondria. Anat Rec 114(3):427–451
19. Palade GE, Porter KR (1954) Studies on the endoplasmic reticulum. I. Its identification in cells in situ. J Exp Med 100(6):641–656
20. Dalton AJ, Felix MD (1954) Cytologic and cytochemical characteristics of the Golgi substance of epithelial cells of the epididymis in situ, in homogenates and after isolation. Am J Anat 94(2):171–207
21. Armstrong PB (1970) A fine structural study of adhesive cell junctions in heterotypic cell aggregates. J Cell Biol 47(1):197–210
22. Goel SC (1970) Electron microscopic studies on developing cartilage. I. The membrane system related to the synthesis and secretion of extracellular materials. J Embryol Exp Morphol 23(1):169–184
23. Kirschner RH, Rusli M, Martin TE (1977) Characterization of the nuclear envelope, pore complexes, and dense lamina of mouse liver nuclei by high resolution scanning electron microscopy. J Cell Biol 72(1):118–132
24. Huxley HE (1957) The double array of filaments in cross-striated muscle. J Biophys Biochem Cytol 3(5):631–648
25. Palay SL, Palade GE (1955) The fine structure of neurons. J Biophys Biochem Cytol 1 (1):69–88
26. Peters A, Palay SL, Webster HD (1991) The fine structure of the nervous system, 3rd edn. Oxford University Press, New York

27. Gray EG (1959) Axo-somatic and axo-dendritic synapses of the cerebral cortex: an electron microscope study. J Anat 93:420–433

28. Torrealba F, Carrasco MA (2004) A review on electron microscopy and neurotransmitter systems. Brain Res Brain Res Rev 47 (1–3):5–17

29. Zhang J, Muller JF, McDonald AJ (2013) Noradrenergic innervation of pyramidal cells in the rat basolateral amygdala. Neuroscience 228:395–408

30. Muller JF, Mascagni F, McDonald AJ (2011) Cholinergic innervation of pyramidal cells and parvalbumin-immunoreactive interneurons in the rat basolateral amygdala. J Comp Neurol 519(4):790–805

31. Nitecka L, Frotscher M (1989) Organization and synaptic interconnections of GABAergic and cholinergic elements in the rat amygdaloid nuclei: single- and double-immunolabeling studies. J Comp Neurol 279(3):470–488

32. Carlsen J, Heimer L (1986) A correlated light and electron microscopic immunocytochemical study of cholinergic terminals and neurons in the rat amygdaloid body with special emphasis on the basolateral amygdaloid nucleus. J Comp Neurol 244(1):121–136

33. Buma P, Roubos EW (1986) Ultrastructural demonstration of nonsynaptic release sites in the central nervous system of the snail Lymnaea stagnalis, the insect Periplaneta americana, and the rat. Neuroscience 17 (3):867–879

34. Zhu PC, Thureson-Klein A, Klein RL (1986) Exocytosis from large dense cored vesicles outside the active synaptic zones of terminals within the trigeminal subnucleus caudalis: a possible mechanism for neuropeptide release. Neuroscience 19(1):43–54

35. Agnati LF et al (1986) A correlation analysis of the regional distribution of central enkephalin and beta-endorphin immunoreactive terminals and of opiate receptors in adult and old male rats. Evidence for the existence of two main types of communication in the central nervous system: the volume transmission and the wiring transmission. Acta Physiol Scand 128(2):201–207

36. Van Bockstaele EJ et al (1996) Ultrastructural evidence for prominent distribution of the mu-opioid receptor at extrasynaptic sites on noradrenergic dendrites in the rat nucleus locus coeruleus. J Neurosci 16 (16):5037–5048

37. Fuxe K et al (2015) Volume transmission in central dopamine and noradrenaline neurons and its astroglial targets. Neurochem Res 40 (12):2600–2614

38. Lee A, Rosin DL, Van Bockstaele EJ (1998) alpha2A-adrenergic receptors in the rat nucleus locus coeruleus: subcellular localization in catecholaminergic dendrites, astrocytes, and presynaptic axon terminals. Brain Res 795(1–2):157–169

39. Steward O, Falk PM, Torre ER (1996) Ultrastructural basis for gene expression at the synapse: synapse-associated polyribosome complexes. J Neurocytol 25(12):717–734

40. Tarrant SB, Routtenberg A (1979) Postsynaptic membrane and spine apparatus: proximity in dendritic spines. Neurosci Lett 11 (3):289–294

41. Tiedge H, Brosius J (1996) Translational machinery in dendrites of hippocampal neurons in culture. J Neurosci 16 (22):7171–7181

42. Pierce JP, van Leyen K, McCarthy JB (2000) Translocation machinery for synthesis of integral membrane and secretory proteins in dendritic spines. Nat Neurosci 3(4):311–313

43. Frotscher M (1992) Application of the Golgi/electron microscopy technique for cell identification in immunocytochemical, retrograde labeling, and developmental studies of hippocampal neurons. Microsc Res Tech 23 (4):306–323

44. Fairen A (2005) Pioneering a golden age of cerebral microcircuits: the births of the combined Golgi-electron microscope methods. Neuroscience 136(3):607–614

45. Blackstad TW (1965) Mapping of experimental axon degeneration by electron microscopy of Golgi preparations. Z Zellforsch Mikrosk Anat 67(6):819–834

46. Stell WK (1965) Correlation of retinal cytoarchitecture and ultrastructure in Golgi preparations. Anat Rec 153(4):389–397

47. Blackstad TW (1975) Electron microscopy of experimental axonal degeneration in photochemically modified Golgi preparations: a procedure for precise mapping of nervous connections. Brain Res 95(2–3):191–210

48. Fairen A, Peters A, Saldanha J (1977) A new procedure for examining Golgi impregnated neurons by light and electron microscopy. J Neurocytol 6(3):311–337

49. Freund TF, Somogyi P (1983) The section-Golgi impregnation procedure. 1. Description of the method and its combination with histochemistry after intracellular

iontophoresis or retrograde transport of horseradish peroxidase. Neuroscience 9 (3):463–474

50. Somogyi P et al (1983) The section-Golgi impregnation procedure. 2. Immunocytochemical demonstration of glutamate decarboxylase in Golgi-impregnated neurons and in their afferent synaptic boutons in the visual cortex of the cat. Neuroscience 9(3):475–490

51. Diana M, Spiga S, Acquas E (2006) Persistent and reversible morphine withdrawal-induced morphological changes in the nucleus accumbens. Ann N Y Acad Sci 1074:446–457

52. Pilati N et al (2008) A rapid method combining Golgi and Nissl staining to study neuronal morphology and cytoarchitecture. J Histochem Cytochem 56(6):539–550

53. Pinto L et al (2012) Immuno-Golgi as a tool for analyzing neuronal 3D-dendritic structure in phenotypically characterized neurons. PLoS One 7(3):e33114

54. Spiga S et al (2011) Simultaneous Golgi-Cox and immunofluorescence using confocal microscopy. Brain Struct Funct 216 (3):171–182

55. Robinson TE, Kolb B (1999) Alterations in the morphology of dendrites and dendritic spines in the nucleus accumbens and prefrontal cortex following repeated treatment with amphetamine or cocaine. Eur J Neurosci 11 (5):1598–1604

56. Carvalho AF et al (2016) Repeated administration of a synthetic cannabinoid receptor agonist differentially affects cortical and accumbal neuronal morphology in adolescent and adult rats. Brain Struct Funct 221 (1):407–419

57. Falowski SM et al (2011) An evaluation of neuroplasticity and behavior after deep brain stimulation of the nucleus accumbens in an animal model of depression. Neurosurgery 69(6):1281–1290

58. Yalow RS, Berson SA (1960) Immunoassay of endogenous plasma insulin in man. J Clin Invest 39:1157–1175

59. von Baumgarten F, Baumgarten HG, Schlossberger HG (1980) The disposition of intraventricularly injected 14C-5,6-DHT-melanin in, and possible routes of elimination from the rat CNS. An autoradiographic study. Cell Tissue Res 212(2):279–294

60. Ryals BM, Westbrook EW (1994) TEM analysis of neural terminals on autoradiographically identified regenerated hair cells. Hear Res 72(1–2):81–88

61. Porstmann B et al (1985) Which of the commonly used marker enzymes gives the best results in colorimetric and fluorimetric enzyme immunoassays: horseradish peroxidase, alkaline phosphatase or beta-galactosidase? J Immunol Methods 79 (1):27–37

62. Krieg R, Halbhuber KJ (2010) Detection of endogenous and immuno-bound peroxidase—the status quo in histochemistry. Prog Histochem Cytochem 45(2):81–139

63. Trojanowski JQ, Obrocka MA, Lee VM (1983) A comparison of eight different chromogen protocols for the demonstration of immunoreactive neurofilaments or glial filaments in rat cerebellum using the peroxidase-antiperoxidase method and monoclonal antibodies. J Histochem Cytochem 31(10):1217–1223

64. Giepmans BN et al (2005) Correlated light and electron microscopic imaging of multiple endogenous proteins using Quantum dots. Nat Methods 2(10):743–749

65. Maranto AR (1982) Neuronal mapping: a photooxidation reaction makes Lucifer yellow useful for electron microscopy. Science 217 (4563):953–955

66. Lubke J (1993) Photoconversion of diaminobenzidine with different fluorescent neuronal markers into a light and electron microscopic dense reaction product. Microsc Res Tech 24 (1):2–14

67. von Bartheld CS, Cunningham DE, Rubel EW (1990) Neuronal tracing with DiI: decalcification, cryosectioning, and photoconversion for light and electron microscopic analysis. J Histochem Cytochem 38 (5):725–733

68. Sandell JH, Masland RH (1988) Photoconversion of some fluorescent markers to a diaminobenzidine product. J Histochem Cytochem 36(5):555–559

69. Singer SJ (1959) Preparation of an electron-dense antibody conjugate. Nature 183 (4674):1523–1524

70. Feldherr CM, Marshall JM Jr (1962) The use of colloidal gold for studies of intracellular exchanges in the ameba Chaos chaos. J Cell Biol 12:640–645

71. Faulk WP, Taylor GM (1971) An immunocolloid method for the electron microscope. Immunochemistry 8(11):1081–1083

72. Lee A, Rosin DL, Van Bockstaele EJ (1998) Ultrastructural evidence for prominent postsynaptic localization of alpha2C-adrenergic receptors in catecholaminergic dendrites in the rat nucleus locus coeruleus. J Comp Neurol 394(2):218–229

73. Reyes BA et al (2014) Using high resolution imaging to determine trafficking of corticotropin-releasing factor receptors in noradrenergic neurons of the rat locus coeruleus. Life Sci 112(1–2):2–9

74. LaVail JH, LaVail MM (1972) Retrograde axonal transport in the central nervous system. Science 176(4042):1416–1417

75. Aldes LD, Boone TB (1984) A combined flat-embedding, HRP histochemical method for correlative light and electron microscopic study of single neurons. J Neurosci Res 11 (1):27–34

76. Mesulam MM (1978) Tetramethyl benzidine for horseradish peroxidase neurohistochemistry: a non-carcinogenic blue reaction product with superior sensitivity for visualizing neural afferents and efferents. J Histochem Cytochem 26(2):106–117

77. Carson KA, Mesulam MM (1982) Electron microscopic demonstration of neural connections using horseradish peroxidase: a comparison of the tetramethylbenzidine procedure with seven other histochemical methods. J Histochem Cytochem 30(5):425–435

78. Carson KA, Mesulam MM (1982) Ultrastructural evidence in mice that transganglionically transported horseradish peroxidase-wheat germ agglutinin conjugate reaches the intraspinal terminations of sensory neurons. Neurosci Lett 29(3):201–206

79. Kravets JL et al (2015) Direct targeting of peptidergic amygdalar neurons by noradrenergic afferents: linking stress-integrative circuitry. Brain Struct Funct 220(1):541–558

80. Reyes BA et al (2011) Amygdalar peptidergic circuits regulating noradrenergic locus coeruleus neurons: linking limbic and arousal centers. Exp Neurol 230(1):96–105

81. Van Bockstaele EJ et al (1991) Subregions of the periaqueductal gray topographically innervate the rostral ventral medulla in the rat. J Comp Neurol 309(3):305–327

82. Schmued LC, Fallon JH (1986) Fluoro-Gold: a new fluorescent retrograde axonal tracer with numerous unique properties. Brain Res 377(1):147–154

83. Gonatas NK et al (1979) Superior sensitivity of conjugates of horseradish peroxidase with wheat germ agglutinin for studies of retrograde axonal transport. J Histochem Cytochem 27(3):728–734

84. Schwab ME, Javoy-Agid F, Agid Y (1978) Labeled wheat germ agglutinin (WGA) as a new, highly sensitive retrograde tracer in the rat brain hippocampal system. Brain Res 152 (1):145–150

85. Prujansky A, Ravid A, Sharon N (1978) Cooperativity of lectin binding to lymphocytes, and its relevance to mitogenic stimulation. Biochim Biophys Acta 508(1):137–146

86. Gerfen CR, Sawchenko PE (1984) An anterograde neuroanatomical tracing method that shows the detailed morphology of neurons, their axons and terminals: immunohistochemical localization of an axonally transported plant lectin, Phaseolus vulgaris leucoagglutinin (PHA-L). Brain Res 290 (2):219–238

87. Reyes BA et al (2005) Hypothalamic projections to locus coeruleus neurons in rat brain. Eur J Neurosci 22(1):93–106

88. Pickel VM et al (1996) GABAergic neurons in rat nuclei of solitary tracts receive inhibitory-type synapses from amygdaloid efferents lacking detectable GABA-immunoreactivity. J Neurosci Res 44(5):446–458

89. Wouterlood FG, Jorritsma-Byham B (1993) The anterograde neuroanatomical tracer biotinylated dextran-amine: comparison with the tracer Phaseolus vulgaris-leucoagglutinin in preparations for electron microscopy. J Neurosci Methods 48(1–2):75–87

90. Veenman CL, Reiner A, Honig MG (1992) Biotinylated dextran amine as an anterograde tracer for single- and double-labeling studies. J Neurosci Methods 41(3):239–254

91. McGovern AE et al (2012) Anterograde neuronal circuit tracing using a genetically modified herpes simplex virus expressing EGFP. J Neurosci Methods 209(1):158–167

92. Kelly RM, Strick PL (2000) Rabies as a transneuronal tracer of circuits in the central nervous system. J Neurosci Methods 103 (1):63–71

93. Chamberlin NL et al (1998) Recombinant adeno-associated virus vector: use for transgene expression and anterograde tract tracing in the CNS. Brain Res 793(1–2):169–175

94. Deller T, Naumann T, Frotscher M (2000) Retrograde and anterograde tracing combined with transmitter identification and electron microscopy. J Neurosci Methods 103 (1):117–126

95. Kohler C, Chan-Palay V, Wu JY (1984) Septal neurons containing glutamic acid decarboxylase immunoreactivity project to the hippocampal region in the rat brain. Anat Embryol (Berl) 169(1):41–44

96. Germroth P, Schwerdtfeger WK, Buhl EH (1989) GABAergic neurons in the entorhinal cortex project to the hippocampus. Brain Res 494(1):187–192

97. Van Bockstaele EJ, Pickel VM (1995) GABA-containing neurons in the ventral tegmental area project to the nucleus accumbens in rat brain. Brain Res 682(1–2):215–221

98. Carr DB, Sesack SR (2000) Projections from the rat prefrontal cortex to the ventral tegmental area: target specificity in the synaptic associations with mesoaccumbens and mesocortical neurons. J Neurosci 20 (10):3864–3873

99. Lubin M, Leonard CS, Aoki C (1998) Preservation of ultrastructure and antigenicity for EM immunocytochemistry following intracellular recording and labeling of single cortical neurons in brain slices. J Neurosci Methods 81(1–2):91–102

100. Kaneko T et al (2000) Predominant information transfer from layer III pyramidal neurons to corticospinal neurons. J Comp Neurol 423 (1):52–65

101. Ralston HJ III et al (1984) Morphology and synaptic relationships of physiologically identified low-threshold dorsal root axons stained with intra-axonal horseradish peroxidase in the cat and monkey. J Neurophysiol 51 (4):777–792

102. Wilson JR, Friedlander MJ, Sherman SM (1984) Fine structural morphology of identified X- and Y-cells in the cat's lateral geniculate nucleus. Proc R Soc Lond B Biol Sci 221 (1225):411–436

103. Tasker JG, Hoffman NW, Dudek FE (1991) Comparison of three intracellular markers for combined electrophysiological, morphological and immunohistochemical analyses. J Neurosci Methods 38(2–3):129–143

104. Branchereau P et al (1996) Pyramidal neurons in rat prefrontal cortex show a complex synaptic response to single electrical stimulation of the locus coeruleus region: evidence for antidromic activation and GABAergic inhibition using in vivo intracellular recording and electron microscopy. Synapse 22 (4):313–331

105. Hamos JE (1990) Synaptic circuitry identified by intracellular labeling with horseradish peroxidase. J Electron Microsc Tech 15 (4):369–376

106. Pickel VM et al (2006) Dopamine D1 receptors co-distribute with N-methyl-D-aspartic acid type-1 subunits and modulate synaptically-evoked N-methyl-D-aspartic acid currents in rat basolateral amygdala. Neuroscience 142(3):671–690

107. Branchereau P et al (1995) Ultrastructural characterization of neurons recorded intracellularly in vivo and injected with lucifer yellow:

108. Cowan RL et al (1994) Analysis of synaptic inputs and targets of physiologically characterized neurons in rat frontal cortex: combined in vivo intracellular recording and immunolabeling. Synapse 17(2):101–114

109. Sabatini DD, Bensch K, Barrnett RJ (1963) Cytochemistry and electron microscopy. The preservation of cellular ultrastructure and enzymatic activity by aldehyde fixation. J Cell Biol 17:19–58

110. Saito T, Keino H (1976) Acrolein as a fixative for enzyme cytochemistry. J Histochem Cytochem 24(12):1258–1269

111. Connolly CN et al (1994) Transport into and out of the Golgi complex studied by transfecting cells with cDNAs encoding horseradish peroxidase. J Cell Biol 127(3):641–652

112. Shu X et al (2011) A genetically encoded tag for correlated light and electron microscopy of intact cells, tissues, and organisms. PLoS Biol 9(4):e1001041

113. Martell JD et al (2012) Engineered ascorbate peroxidase as a genetically encoded reporter for electron microscopy. Nat Biotechnol 30 (11):1143–1148

114. Bishop D et al (2011) Near-infrared branding efficiently correlates light and electron microscopy. Nat Methods 8(7):568–570

115. Parsons DF et al (1974) Electron microscopy and diffraction of wet unstained and unfixed biological objects. Adv Biol Med Phys 15:161–270

116. Dubochet J (2012) Cryo-EM—the first thirty years. J Microsc 245(3):221–224

117. Mayer E, Brüggeller P (1980) Complete vitrification in pure liquid water and dilute aqueous solutions. Nature 288:569–571

118. Dubochet J, McDowall AW (1981) Vitrification of pure water for electron microscopy. J Microsc 124:RP3–RP4

119. Dubochet J et al (1988) Cryo-electron microscopy of vitrified specimens. Q Rev Biophys 21(2):129–228

120. Al-Amoudi A et al (2004) Cryo-electron microscopy of vitreous sections. EMBO J 23 (18):3583–3588

121. McDonald KL, Auer M (2006) High-pressure freezing, cellular tomography, and structural cell biology. Biotechniques 41 (2):137, 139, 141 passim

122. Korogod N, Petersen C, Knott GW (2015) Ultrastructural analysis of adult mouse

neocortex comparing aldehyde perfusion with cryo fixation. Elife 4

123. Studer D et al (2014) Capture of activity-induced ultrastructural changes at synapses by high-pressure freezing of brain tissue. Nat Protoc 9(6):1480–1495

124. White EL, Hersch SM (1982) A quantitative study of thalamocortical and other synapses involving the apical dendrites of corticothalamic projection cells in mouse SmI cortex. J Neurocytol 11(1):137–157

125. Hayworth KJ et al (2014) Imaging ATUM ultrathin section libraries with Wafer-Mapper: a multi-scale approach to EM reconstruction of neural circuits. Front Neural Circuits 8:68

126. Kevin LG (2006) Comment on "Effects of short vs. prolonged mechanical ventilation on antioxidant systems in piglet diaphragm" by Jaber et al. Intensive Care Med 32 (9):1446, author reply 1447

127. Denk W, Horstmann H (2004) Serial blockface scanning electron microscopy to reconstruct three-dimensional tissue nanostructure. PLoS Biol 2(11):e329

128. Wanner AA, Kirschmann MA, Genoud C (2015) Challenges of microtome-based serial-block-face scanning electron microscopy in neuroscience. J Microsc 259(2):137–142

129. Mikula S, Binding J, Denk W (2012) Staining and embedding the whole mouse brain for electron microscopy. Nat Methods 9 (12):1198–1201

130. Mikula S, Denk W (2015) High-resolution whole-brain staining for electron microscopic circuit reconstruction. Nat Methods 12 (6):541–546

131. Nogales E, Scheres SH (2015) Cryo-EM: a unique tool for the visualization of macromolecular complexity. Mol Cell 58(4):677–689

132. Fernandez-Busnadiego R et al (2011) Insights into the molecular organization of the neuron by cryo-electron tomography. J Electron Microsc (Tokyo) 60(Suppl 1): S137–S148

133. Lucic V et al (2007) Multiscale imaging of neurons grown in culture: from light microscopy to cryo-electron tomography. J Struct Biol 160(2):146–156

134. Shahmoradian SH et al (2014) Preparation of primary neurons for visualizing neurites in a frozen-hydrated state using cryo-electron tomography. J Vis Exp 84:e50783

135. Alzheimer's Association (2015) 2015 Alzheimer's disease facts and figures. ALzheimers Dement 11:332–384

136. Clare R et al (2010) Synapse loss in dementias. J Neurosci Res 88(10):2083–2090

137. Harris KM et al (2015) A resource from 3D electron microscopy of hippocampal neuropil for user training and tool development. Sci Data 2:150046

138. Alonso-Nanclares L et al (2013) Synaptic changes in the dentate gyrus of APP/PS1 transgenic mice revealed by electron microscopy. J Neuropathol Exp Neurol 72(5):386–395

139. Kuwajima M, Spacek J, Harris KM (2013) Beyond counts and shapes: studying pathology of dendritic spines in the context of the surrounding neuropil through serial section electron microscopy. Neuroscience 251:75–89

140. Nuntagij P et al (2009) Amyloid deposits show complexity and intimate spatial relationship with dendrosomatic plasma membranes: an electron microscopic 3D reconstruction analysis in 3xTg-AD mice and aged canines. J Alzheimers Dis 16(2):315–323

141. Fiala JC et al (2007) Mitochondrial degeneration in dystrophic neurites of senile plaques may lead to extracellular deposition of fine filaments. Brain Struct Funct 212 (2):195–207

Neuromethods (2016) 115: 21–33
DOI 10.1007/7657_2015_78
© Springer Science+Business Media New York 2015
Published online: 18 July 2015

Dual-Labeling Immuno-Electron Microscopic Method for Identifying Pre- and Postsynaptic Profiles in Mammalian Brain

Xiao-Bo Liu and Hwai-Jong Cheng

Abstract

Mammalian nervous system communicates mainly through chemical synapses with neurotransmitters in the presynaptic axon terminal and their receptors on the postsynaptic dendritic targets. We describe here an immuno-electron microscopic (immuno-EM) method to simultaneously label presynaptic neurotransmitters with postembedding immunogold and postsynaptic components with preembedding immunoperoxidase. This dual-immuno-EM labeling technique allows reliable distinction of the two types of EM signals in closely related ultrastructures. The key steps of the procedure and the detailed protocol are provided.

Keywords Immuno-EM, Dual labeling, Excitatory neurotransmitter, Immunoperoxidase, Gold particles, Postsynaptic density

---

1 Introduction

Neurons in the nervous system connect with each other through a functional contact site called synapse [1, 2]. In a typical excitatory synapse, the principal neurotransmitter glutamate is contained in the synaptic vesicles inside the presynaptic axon terminal. When the terminal is depolarized, the synaptic vesicles are fused to the membrane and the glutamate molecules are released to the synaptic cleft. The glutamate molecules then bind to glutamate receptors on the membrane of postsynaptic spine or dendrite. For ionotropic glutamate receptors, the binding of glutamate opens ion channels, allowing positively charged ions (Na^+ and Ca^{2+}) to flow through, which leads to depolarization of the membrane potential and subsequent changes in the postsynaptic site [3, 4]. In addition, the glutamate itself is an important metabolic component in the axon terminals. It plays an essential role in intermediary metabolism and is required for maintaining the basic function of a neuron [5, 6].

The physiological connection of glutamate-mediated excitatory synapse in the brain has been intensively studied in the past several decades. How the release of glutamate or other related molecules regulates the physiological response in the postsynaptic site has been well characterized. The biochemical pathways regulating the synthesis and transport of glutamate in the presynaptic axon terminal have also been well documented [4]. However, the subcellular distribution of glutamate in the neuron and its axon terminal has been difficult to study due to a lack of specific tool to identify the molecule. In addition, to understand the ultrastructure-function relationship in a glutamate-mediated excitatory synapse, one needs to use not only a specific marker to locate the subcellular location of glutamate in the presynaptic axon terminal, but also another marker to identify the postsynaptic components at the same time.

There have been several challenges to use electron microscopic (EM) techniques to simultaneously detect the ultrastructural relationship of the pre- and postsynaptic components before. The main reasons are the following: (a) The presynaptic component in the axon terminal and the postsynaptic element in the spine or dendrite have to be differentiated by two distinct signals under EM. (b) All the pre- and postsynaptic ultrastructures have to be well preserved in the process of preparation for EM analysis. Thanks to the development of new and highly specific antibodies for glutamate and other pre- and postsynaptic proteins in the last two decades, numerous studies have been carried out aiming at identifying the pre- and postsynaptic elements using immuno-EM methods [5–13]. The immuno-EM techniques have been improved significantly throughout these years. Recently, double-labeling immuno-EM techniques have been developed to simultaneously label presynaptic glutamate and postsynaptic elements in excitatory synapses [13–17]. Here we provide a detailed double-labeling immuno-EM protocol. Using glutamatergic excitatory neurons in the cerebral cortex and thalamus as an example [15–17], we describe the EM preparation procedures for identifying the glutamate in the presynaptic terminal by immunogold labeling and the postsynaptic CaM Kinase-α and ionotropic glutamate receptor subunits by immunoperoxidase labeling.

2 Materials

2.1 Reagents

2.1.1 Fixation and Buffer Solutions

4 % Paraformaldehyde and 0.5 % glutaraldehyde in 0.1 M phosphate buffer (PB), pH 7.4: The fixative should be freshly prepared. Add 16 g of paraformaldehyde powder into preheated 100 ml distilled deionized (d.d.) H_2O at 65 °C in a beaker. Add two drops of 5 M sodium hydroxide and a magnetic bar into the beaker, place it on a heated stirrer, and stir until the solution clears (about 2 min.). Add the cleared solution to 200 ml of cold 0.2 M PB and then add 8 ml of freshly opened glutaraldehyde (25 % vol/vol aqueous solution).

Adjust pH to 7.4 and then make the final volume to 400 ml with d. d. H_2O. Leave the solution to cool to 20 °C.

2 % Osmium tetroxide in 0.1 M PB: Add 4 ml of 4 % aqueous osmium tetroxide solution to 4 ml d.d. H_2O and then add 8 ml 0.2 M PB. The solution should be freshly made and stored in cool. *Uranyl acetate (1 % wt/vol in d.d. H_2O):* Add 0.5 g of uranyl acetate powder to 50 ml of d.d. H_2O. Agitate the solution gently until it is completely dissolved.

Tris-buffered saline (0.05 M TBS, pH 7.6 or pH 8.2).

Phosphate buffer (PB, pH 7.4, 0.1 M or 0.2 M).

2.1.2 Immuno-cytochemistry	*Primary antibodies:*

Anti-CaM Kinase-α (mouse monoclonal antibody, Chemicon, MAB8699) (formerly Roche Catalog Number 1481703).

Anti-glutamate (polyclonal antibody raised in rabbit, Sigma, product # G6642).

Anti-GABA (polyclonal antibody raised in rabbit, Sigma, product # A2052).

Anti-GluR2/3 (anti-GluR2/3 polyclonal antibody raised in rabbit, Chemicon, 07-598).

Secondary antibodies and other supplies:

Horse anti-mouse biotinylated antibody (Chemicon).

Goat anti-rabbit biotinylated antibody (Chemicon).

Gold particle-conjugated goat anti-rabbit antibody (10 or 15 nm gold particles, British BioCell/Ted Pella).

Avidin-biotin complex (Vectastain, Vectorlabs, cat. No. PK-6100).

3,3′-Diaminobenzidine tetrahydrochloridehydrate (DAB) (Sigma-Aldrich, cat. No. D5637-1G).

2.1.3 EM Embedding   Araldite 502 kit (EMS).

2.2 Equipments

2.2.1 EM Embedding

Glass histology slides and cover slips.

Mold-separating agent (Sigmacote).

Oven (60 °C).

2.2.2 Sample Block Preparation

Cyanoacrylate glue.

Sharp surgical blade.

Dissecting microscope.

Glass knife maker.

Light microscope.

Ultramicrotome (Leica UCT).

2.2.3 Thin Sectioning and Staining	Diamond knife (45°).
	Grid storage box.
	Single-slot grids with formvar support film and carbon coated; slot dimensions, 2 mm × 1 mm.
	Ultramicrotome (Leica UCT).
2.2.4 Imaging	TEM with digital camera (EM: Philips CM120, Gatan 2 k × 2 k CCD camera and digital micrograph software).
	Computer.
	Photoshop software.

3   Experimental Approach

The dual-labeling immuno-EM technique is developed to study the relationship of two molecules in closely related ultrastructures. To investigate the subcellular relationship of molecules in an excitatory synapse at EM level, the ultrastructural features of the synapse have to be well preserved and the dual-molecular EM signals need to be simultaneously displayed and differentiated. We have previously established a dual-labeling immuno-EM method to study the relationship between the excitatory presynaptic neurotransmitter and the postsynaptic elements in the thalamus and cerebral cortex of cat and rodent. Outlined here are the major steps for the method, followed by the detailed procedures.

1. *Animal brain tissues:* The adult cat or rat is perfused with fixatives under deep anesthesia. The brain is then removed, vibratome-sectioned at 60–80 μm, and stored in 0.1 M PB.

2. *Preembedding immunoperoxidase staining:* This step is to visualize the postsynaptic neuronal components. It is therefore aimed to label the dendritic processes and cell bodies of neurons that are targeted by excitatory axon terminals in the thalamus or cerebral cortex. For immuno-EM, CaM Kinase-α is an ideal marker for thalamic relay cells and cortical pyramidal neurons. Other useful markers are ionotropic glutamate receptor subunits (GluR2/3 subunits). The prepared vibratome sections are processed for immunoperoxidase staining for the markers. Since triton (or other similar detergents) is known to compromise membrane ultrastructure, it is important to note that triton should be avoided in all steps for EM immunoperoxidase staining. The intensity of the immunostaining signal for immuno-EM is overall reduced compared to that of the conventional immunostaining for light microscopy in which triton is commonly used. After immunostaining, the wet sections are examined under light microscope and the sections

containing immunolabeled neurons are selected for further EM embedding.

3. *Electron microscopy embedding and sectioning:* The selected immunolabeled vibratome sections are osmicated, dehydrated in graded ethanol and acetone, and flat-embedded into Araldite. After polymerization, embedded sections are further examined under the light microscope. These flat-embedded sections can be permanently saved. For further EM analysis, these embedded sections can be examined at much higher magnification to identify the labeled neurons and their fine dendritic processes. The identified area can be dissected out for thin sectioning. The thin sections can then be placed on grids for EM examination.

4. *Electron microscopic examination of preembedding immuno-peroxidase-labeled thin sections:* The immunoperoxidase-labeled neurons and their dendritic processes can now be visualized under the EM. The quality of the ultrastructure and the intensity of the immunoperoxidase staining on the section are evaluated for postembedding immunogold labeling. In general, the plasma membrane structure should be well preserved and the postsynaptic markers should be moderately labeled with electron-dense materials.

5. *Postembedding immunogold labeling for glutamate transmitter:* The postembedding immunogold staining is applied to the preembedding immunoperoxidase-labeled thin sections for identifying the presynaptic inputs. At this step, the antibodies against neurotransmitters such as glutamate (or GABA for inhibitory synapse) can be used for immunogold staining. The single slots or mesh grids containing preembedding immunoperoxidase-labeled sections are incubated with solutions containing anti-glutamate antibody for several hours. After extensive wash, the grids are incubated in solutions containing secondary antibody conjugated with either 10, 15, or 20 nm sized gold particles. After another extensive wash, the grids are stained with uranyl acetate and lead citrate, and then examined under the EM.

Examples of the dual-labeling immuno-EM are demonstrated in Figs. 1 and 2. In these figures, either the excitatory neurotransmitter glutamate (Fig. 1a, b and 2a) or the inhibitory neurotransmitter GABA (Fig. 2b) is labeled with gold particles, which locate within axon terminals. The postsynaptic CAMKII-α (arrows in Fig. 1) or GluR2/3 (arrows in Fig. 2) markers for excitatory synapses are labeled with electron-dense immunoperoxidase reaction products in the postsynaptic densities. In Fig. 1, the dual labeling is clearly confirmed in serial sections. In Fig. 2, the immunogold staining for glutamate (Fig. 2a) or GABA (Fig. 2b) in the dual labeling in serial sections demonstrates the association of glutamatergic (t) axon terminal with immunoperoxidase-labeled

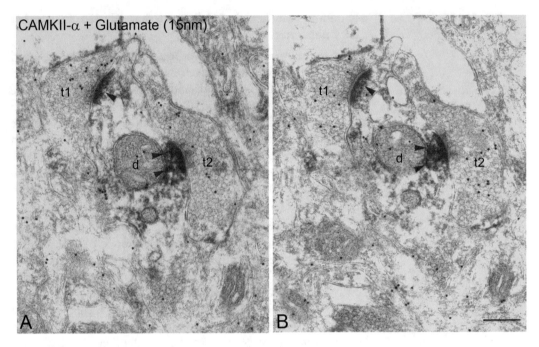

Fig. 1 An example of dual-labeling immuno-EM method to study the ultrastructural relationship of the presynaptic glutamate and postsynaptic CaMKII-α in the ventral posterior nucleus of rat thalamus. Serial immuno-EM micrographs (**a** and **b**) show that the presynaptic glutamate-containing corticothalamic axon terminals (t1 and t2; the glutamate is immunogold-labeled with 15 nm gold particles) are closely associated with a postsynaptic dendrite (d), in which the strong CaMKII-α immunoperoxidase reaction products (*arrow heads*) are observed at the postsynaptic sites. Note that the gold particles are consistently present in both terminals in serial sections. Bar = 0.5 μm

GluR2/3 in the postsynaptic density (d), but not the neighboring GABA-positive axon terminal (gt).

Figure 3 summarizes the major steps outlined above.

4   Detailed Procedures

1. *Protocol I: Preembedding immunoperoxidase labeling*

   (a) *Fixation and tissue preparation*

   Animal use should be carried out under the Institutional and National Guidelines. The animal is maintained under deep anesthesia with Nembutal, and perfused through its left ventricle first with saline and then with cold 0.1 M PB (pH 7.4) containing 4 % paraformaldehyde and 0.2–0.5 % glutaraldehyde. The perfusion generally takes about 20 min to complete. The rate of perfusion must not be too fast as some of the small vessels in the vasculature may burst, which would create a low-resistance flow and prevent the

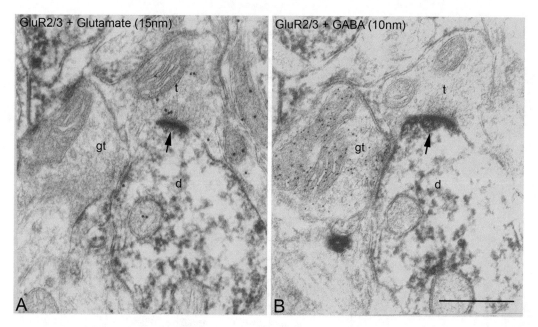

Fig. 2 Another example of dual-labeling immuno-EM method to study the ultrastructural relationship of the presynaptic glutamate (immunogold with 15 nm gold particles) or GABA (immunogold with 10 nm gold particles) and postsynaptic GluR2/3 subunit (immunoperoxidase labeling, *arrows*) in the ventral posterior nucleus of rat thalamus. Serial immuno-EM sections were processed with different immunogold labeling for glutamate (**a**) and GABA (**b**), respectively. The micrograph in (**a**) shows that the glutamate-containing corticothalamic axon terminal (t, labeled with 15 nm gold particles) is closely associated with a postsynaptic density (*arrow*) labeled with strong GluR2/3 subunit immunoperoxidase reaction products in a dendrite (d). Note that an adjacent axon terminal (gt) is absent of glutamate-labeled gold particles. The serial micrograph in (**b**) shows that this axon terminal (gt) is immunogold labeled with GABA (10 nm gold particles). Bar = 0.5 μm

delivery of fixatives to the brain. A good quality of perfusion quickly turns the liver into light brown color within few seconds after the perfusion starts. Remove the brain from the skull and postfix it in 4 % paraformaldehyde (in 0.1 M PB) for 3 h at room temperature or overnight at 4 °C.

(b) *Vibratome sectioning*

Glue the brain block to the vibratome stage with cyanoacrylate and cut the brain into sections at 60–80 μm in thickness. The sections are collected in cold 0.1 M PB and stored at 4 °C.

(c) *Immunocytochemical staining*

– Wash the sections in 0.1 M PB three times (3×) with 5 min each; incubate in 5–10 % normal serum in 0.1 M PB for 1 h at room temperature (RT) on a rotator.

– Incubate in primary antibody with 5 % normal serum for 48 h at 4 °C.

– Wash in 0.1 M PB 3× with 5 min each at RT.

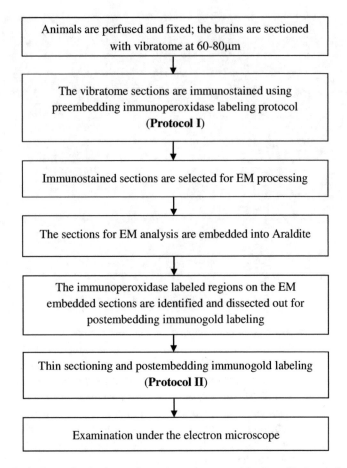

Fig. 3 A flow chart shows key steps in preembbeding (Protocol I) and postembedding (Protocol II) labeling procedures

- Incubate in secondary antibody (1:150 to 1:250) in 0.1 M PB for 1.5 h at RT.
- Wash in 0.1 M PB 3× with 5 min each at RT.
- Incubate in avidin–biotin complex solution (2 μl solution A + 2 μl solution B in 10 ml 0.1 M PB) for 1.5 h at RT.
- Wash in 0.1 M PB 3× with 5 min each.
- Incubate in 0.001 % DAB/0.1 M PB solution for 10 min, add 0.015 % (vol/vol) H_2O_2 to the solution, and incubate for 5–10 min. The immunoperoxidase signals are carefully monitored under light microscope to prevent overdevelopment of the signal in the background.
- Wash sections in 0.1 M PB 3× with 5 min each.

(d) *Section embedding for electron microscopy*

- The sections containing immunostained neurons or their processes are selected for EM embedding. Wash the selected sections in 0.1 M PB 3× with 5 min each at RT.

- Postfix in 1–2 % OsO_4 in 0.1 M PB (freshly made) for 20 min.

- Wash 3× in d.d. H_2O.

- Dehydrate the sections in graded alcohol series (2× in 70 %, 2× in 90 %, 2× in 95 %, and 2× in 100 %, with 5 min each). Special care should be taken not to allow the sections to air-dry at any time.

- Dehydrate in 100 % aqueous acetone 2× with 5 min each.

- Place the sections in freshly made 1:1 acetone/Araldite solution for 1 h on the rotator.

- Replace the sections with pure Araldite and leave them at 4 °C overnight.

- Replace with fresh Araldite on the next day and rotate for 4 h at RT.

- Carefully remove each section from the resin with wooden cocktail sticks, place them on a glass microscopic slide that has been coated with Sigmacote (Sigma), and then cover them with a Sigmacote-pretreated cover slip. There should be no air bubble. Make sure that there is enough resin to fill the gap between the slides. However, too much resin may also cause the sections to drift out of the slides.

- Place the embedded sections in 60 °C oven for 48 h. When the resin has hardened, the glass slides can be separated from the resin with the embedded sections by using a razor blade. The mold-separating agents (Sigmacote) prevent the resin from bonding to the glass slides.

(e) *Preparation of embedded sections for EM examination*

- Embedded sections are examined in the light microscope; the immunolabeled neurons are identified and photographed.

- Select properly immunostained regions for further electron microscopic examinations and dissect the regions out with a razor blade under a dissecting microscope.

- Carefully place the sections that are dissected out on a blank resin block and attach the sections with cyanoacrylate glue.

– Trim the resin block with glass knife on an ultramicrotome.

– Section the block serially with a clean diamond knife. The thin sections are cut at 70 nm. Collect the sections onto formvar-coated single-slot Nickel grids and store them in a plastic grid box.

– Load grids onto a plastic grid supporting plate and then place them into a petri dish.

– Flood the plate and grids with 1 % uranyl acetate, cover the petri dish with a box to block the light, and stain for 10–15 min.

– Wash the grids with d.d. H_2O, dry, and then replace with lead citrate solution. At the same time, place some sodium hydroxide pellets in the petri dish (to remove any carbon dioxide) and cover for 10 min.

– Remove lead citrate and replace with d.d. H_2O, wash with d.d. H_2O 3×, and dry for at least 1 h.

– Remove the grids from the grid-supporting plate and store them in a grid box.

– During EM analysis, identify the immunolabeled profiles. Evaluate the ultrastructural quality and select appropriate regions for further postembedding immunogold staining.

2. *Protocol II: Postembedding immunogold labeling*

(a) *Pretreatment of sections and the general considerations:*
In most cases, ultrathin sections are cut and collected on the grids made from inert metal such as nickel or gold. Copper grids should be avoided because they may succumb during the various treatments. The grids may or may not be supported with formvar film. For serial section analysis, thin sections are usually collected on single-slot formvar-coated nickel grids. The formvar film prevents penetration of antibodies and allows staining of only one side of the thin sections. For routine EM analysis, sections are collected on fine mesh naked grids without formvar film coating. This allows staining on both sides of the thin section by immerging them in the solution.

Another consideration is the pretreatment of Epon- and Araldite-embedded osmicated sections with periodic acid (HIO_4) and sodium metaperiodate ($NaIO_4$) to remove OsO_4. However, it is not necessary to treat nonosmicated blocks with these chemicals before immunostaining. The same principle is applied to Lowicryl K4M sections. The osmicated sections in the immunoperoxidase labeling method generally need to be deosmicated prior to the immunogold staining as described in detail in section b.

(b) *Procedures of postembedding immunogold staining for EM*

– Insert the grids containing thin sections vertically into the slits of a grid support pad such that the sections are covered by solution. Note that for the sections on supporting film, only one side can be stained.

– Place the support pad on a platform in a moisture box (wet paper towel can be used) and cover it with a lid. It is very important to keep the sections moisturized.

– Apply freshly made 1 % $NaIO_4$ solution to the grids for 10 min.

– Wash in d.d. H_2O 2× with 3 min each.

– Apply freshly made 1 % HIO_4 solution to the grids for 10 min.

– Wash in d.d. H_2O 2× with 3 min each.

– Wash in 0.05 M Tris buffer saline (TBS) plus 0.1 % triton (pH 7.6) with 3 min.

– Apply 5 % normal serum in 0.05 M TBS plus 0.1 % triton solution (pH 7.6), 20 min.

– Apply primary antibody diluted in 5 % normal serum in 0.05 M TBS for 4 h at RT.

– Wash in 0.05 M TBS plus 0.1 % triton (pH 7.6), 2× with 3 min each.

– Wash in 0.05 M TBS plus 0.1 % triton (pH 8.2) 2× with 3 min each.

– Apply secondary antibody conjugated to gold particles (1:20 to 1:50 diluted in 1 % normal serum in 0.05 M TBS plus 0.1 % triton, pH 8.2) for 1.5–2 h.

– Wash in 0.05 M TBS (pH 7.6) without triton 2× for 5 min each.

– Wash in d.d. H_2O 2× for 5 min each.

– Dry the sections for EM staining.

– Examine the sections under EM.

5 Important Notes

1. *Balance of ultrastructural quality and immunostaining:*
The ultrastructural details of the tissue are essential for EM analysis. The preservation of the subcellular ultrastructure is often achieved by adding glutaraldehyde in the fixative solution. In this dual-labeling immuno-EM protocol, we use

0.2–0.5 % glutaraldehyde in the preembedding immunoperoxidase labeling. Although this level of glutaraldehyde might be enough for preserving ultrastructures in non-neuronal tissues [18], it is not ideal for preserving the plasma membranes of the neuronal tissues in general. However, higher amount of glutaraldehyde might affect the affinity of antibody to the antigen, compromising the quality of immuno-staining. As seen in Figs. 1 and 2, the neuronal ultrastructures such as synaptic vesicles and mitochondria might not be of the best quality, but the immuno-staining signals are strong and clear. Keep in mind that we add triton (one of the detergents) and sodium metaperiodate (which has been shown to have bleaching effect at high concentration) in the postembedding immunogold labeling protocol. These reagents would further compromise the quality of the ultrastructure. Thus, it is important to test the best concentration of glutaraldehyde used in the dual-labeling immuno-EM protocol. A small increase of glutaraldehyde concentration without compromising the immuno-staining may significantly enhance the quality of the ultrastructure.

2. *Pretreatment of thin sections and preservation of ultrastructure*: Several other chemicals/reagents used in the dual-labeling immuno-EM protocol might also compromise the quality of the ultrastructure for analysis. In postembedding immunogold labeling, thin sections are first treated with sodium metaperiodate solution ($NaIO_4$) to remove osmium tetroxide (OsO_4) from the tissue and to unmask the antigen (marker) sites for immunogold labeling [18, 19]. The treatment of sodium metaperiodate might compromise the ultrastructure and decrease the contrast of EM images. Some other chemicals are also used to treat thin sections and enhance immunolabeling. It was reported that sodium hydroxyl-ethanolic hydrogen peroxide (NaOH/absolute ethyl alcohol) can be used to remove resins from epoxy thin sections [20] and preserve the section quality, but this chemical is too harsh when directly applied to the thin sections. Diluted NaOH/absolute ethyl alcohol may help detect the antibody in resin sections [21, 22]. We have found that prolonged treatment of these two reagents (more than 10 min for $NaIO_4$, or more than 2 min for NaOH/absolute ethyl alcohol) can intensify immunogold labeling but significantly affect the membrane ultrastructure. It is important to optimize the duration of treatment to preserve the structural integrity of the tissue.

References

1. Peters A, Palay SL, Webster HD (1991) The fine structure of the nervous system. Oxford University Press, New York

2. Douglas R, Markram H, Martin K (2004) Neocortex. In: Shepherd GM (ed) The synaptic organization of the brain. Oxford University Press, Oxford, pp 499–558

3. Deutch AY, Roth RH (2003) Neurotransmitters. In: Squire LR, Bloom FE, McConnell SK, Roberts JL, Spitzer NC, Zigmond MJ (eds) Fundamental neuroscience. Academic, Waltham, MA, pp 163–196

4. Dingeldine R, McBain CJ (1994) Excitatory amino acid transmitters. In: Siegel GJ, Agranoff BW, Albers RW, Molinoff PB (eds) Basic neurochemistry. Raven, New York, pp 367–388

5. Bramham CR, Torp R, Zhang N, Storm-Mathisen J, Ottersen OP (1990) Distribution of glutamate-like immunoreactivity in excitatory hippocampal pathways: a semiquantitative electron microscopic study in rats. Neuroscience 39:405–417

6. Ottersen OP, Laake JH, Storm-Mathisen J (1990) Demonstration of a releasable pool of glutamate in cerebellar mossy and parallel fibre terminals by means of light and electron microscopic immunocytochemistry. Arch Ital Biol 128:111–125

7. Azkue J, Bidaurrazaga A, Mateos JM, Sarria R, Streit P, Grandes P (1995) Glutamate-like immunoreactivity in synaptic terminals of the posterior cingulopontine pathway: a light and electron microscopic study in the rabbit. J Chem Neuroanat 9:261–269

8. Montero VM (1990) Quantitative immunogold analysis reveals high glutamate levels in synaptic terminals of retino-geniculate, cortico-geniculate, and geniculo-cortical axons in the cat. Vis Neurosci 4:437–443

9. Ottersen OP (1987) Postembedding light- and electron microscopic immunocytochemistry of amino acids: description of a new model system allowing identical conditions for specificity testing and tissue processing. Exp Brain Res 69:167–174

10. Ottersen OP (1989) Postembedding immunogold labelling of fixed glutamate: an electron microscopic analysis of the relationship between gold particle density and antigen concentration. J Chem Neuroanat 2:57–66

11. Rubio ME, Juiz JM (2004) Differential distribution of synaptic endings containing glutamate, glycine, and GABA in the rat dorsal cochlear nucleus. J Comp Neurol 477:253–272

12. Storm-Mathisen J, Ottersen OP (1990) Immunocytochemistry of glutamate at the synaptic level. J Histochem Cytochem 38:1733–1743

13. Van Bockstaele EJ, Saunders A, Commons KG, Liu XB, Peoples J (2000) Evidence for coexistence of enkephalin and glutamate in axon terminals and cellular sites for functional interactions of their receptors in the rat locus coeruleus. J Comp Neurol 417:103–114

14. De Biasi S, Amadeo A, Spreafico R, Rustioni A (1994) Enrichment of glutamate immunoreactivity in lemniscal terminals in the ventropostero lateral thalamic nucleus of the rat: an immunogold and WGA-HRP study. Anat Rec 240:131–140

15. Liu XB, Jones EG (1996) Localization of alpha type II calcium calmodulin-dependent protein kinase at glutamatergic but not gamma-aminobutyric acid (GABAergic) synapses in thalamus and cerebral cortex. Proc Natl Acad Sci U S A 93:7332–7336

16. Liu XB (1997) Subcellular distribution of AMPA and NMDA receptor subunit immunoreactivity in ventral posterior and reticular nuclei of rat and cat thalamus. J Comp Neurol 388:587–602

17. Liu XB, Jones EG (1997) Alpha isoform of calcium-calmodulin dependent protein kinase II (CAM II kinase-alpha) restricted to excitatory synapses in the CA1 region of rat hippocampus. Neuroreport 8:1475–1479

18. Morris RE, Ciraolo GM (1997) A universal post-embedding protocol for immunogold labelling of osmium-fixed, epoxy resin-embedded tissue. J Electron Microsc (Tokyo) 46:315–319

19. Bendayan M, Zollinger M (1983) Ultrastructural localization of antigenic sites on osmium-fixed tissues applying the protein A-gold technique. J Histochem Cytochem 31:101–109

20. Maxwell MH (1978) Two rapid and simple methods used for the removal of resins from 1.0 μm thick epoxy sections. J Microsc 112:253–255

21. Kuhlmann WD, Peschke P (1982) Advances in ultrastructural postembedment localization of antigens in Epon sections with peroxidase labelled antibodies. Histochemistry 75:151–161

22. Rodning CB, Erlandsen SL, Wilson ID (1980) Immunohistochemical identification of immunoglobulin A on ultrathin tissue sections. Am J Anat 157:221–224

Neuromethods (2016) 115: 35–62
DOI 10.1007/7657_2015_100
© Springer Science+Business Media New York 2016
Published online: 23 February 2016

Analyzing Synaptic Ultrastructure with Serial Section Electron Microscopy

Jennifer N. Bourne

Abstract

Electron microscopy is a powerful tool that has advanced our understanding of neuroanatomy exponentially over the last several decades. In particular, serial section electron microscopy (ssEM) has provided a three-dimensional context in which to examine neuronal cell biology, structural synaptic plasticity, and anatomical connectivity across multiple brain regions. Here we provide methods and procedures for obtaining, imaging, and analyzing serial sections of brain tissue. In addition, examples and descriptions are provided to help identify different types of spines and synapses and subcellular structures such as polyribosomes and smooth endoplasmic reticulum.

Keywords: Three-dimensional reconstructions, Dendritic spines, Synapses, Polyribosomes, Brain slices

1 Background

In the 1950s, transmission electron microscopy (TEM) was first used to reveal the structural diversity of the central nervous system and unequivocally demonstrate that neurons communicate through synaptic junctions [1, 2]. Since then, imaging and reconstructions of neuronal structures through serial thin sections have provided a 3-dimensional context in which to examine the anatomical connectivity of numerous brain regions [3–7] and evaluate the morphological substrates of synaptic plasticity, learning, and sensory input [8–25]. While serial section electron microscopy (ssEM) is a labor-intensive and time-consuming process, the data yielded can be used to address multiple scientific questions [7, 26]. ssEM provides the needed nanometer resolution to localize and measure synaptic connections defined by the postsynaptic density and presynaptic vesicles, and identify ultrastructural substrates of key cellular functions such as sites of local protein synthesis indicated by polyribosomes [9–11] and local regulation of intracellular calcium and trafficking of membrane proteins by the network of smooth endoplasmic reticulum (SER) that extends throughout the neuron [27, 28]. Thus, ssEM is an essential tool for nanoscale analysis of neuronal cell biology and anatomical connectivity.

Recent advances in technology have automated many of the steps associated with ssEM, including serial block-face scanning electron microscopy, automated tape-collecting ultramicrotomes, focused ion-beam scanning electron microscopy, and automated transmission-mode scanning electron microscopy [29–32]. In addition, computer-assisted image analysis to automate alignment and segmentation of objects through ssEM sections are being developed and hold promise for future application [32–34]. Ideally, a block containing neuronal tissue of interest could be sectioned or scanned, serial images collected, and everything within the imaged neuropil completely reconstructed and quantified, all with minimal human interference and bias. Unfortunately, with the exception of transmission-mode scanning electron microscopy [31], many of these techniques lack the nanometer resolution to unequivocally identify smaller structures such as polyribosomes and may not be financially feasible for those who could still benefit from the use of ssEM. Selective reconstruction or analysis of neuronal elements of interest imaged through ssEM can provide essential and unambiguous information even with relatively limited resources. Here we describe methods optimized to preserve neuronal ultrastructure, produce, collect, and image uniform serial sections, and identify spines, synapses, and subcellular organelles.

2 Equipment, Materials, and Setup

2.1 Fixation

While the gold standard for tissue preservation is rapid in vivo perfusion of fixatives, many experimental setups involve preparations where perfusion is not an option. Here we focus on the preparation of acute brain slices, either adhered onto a net from an interface chamber or free-floating from a submersion chamber. Perfusion-quality tissue preservation is routinely obtained by microwave-enhanced rapid fixation followed by routine tissue processing [8, 35, 36]. Optimal results were obtained with a fixative composed of 6 % glutaraldehyde and 2 % paraformaldehyde in 0.1 M sodium cacodylate buffer. All of these chemicals are hazardous and need to be handled in a fume hood with gloves.

1. *Reagents*: 6 % glutaraldehyde, 2 % paraformaldehyde in 0.1 M sodium cacodylate buffer (100 mL total):

   50 mL 0.2 M sodium cacodylate buffer

   12.5 mL 16 % paraformaldehyde

   8.6 mL 70 % glutaraldehyde

   ~28.9 mL ultrapure water (18.2 M$\Omega \cdot$ cm) (Start with less and then use to q.s. to final volume of 100 mL)

2. *Equipment*:

Microwave—either in a fume hood or properly vented

Neon Bulb Array

Water load to absorb reflected irradiation (a load cooler or beaker filled with water will work)

35 mm polystyrene petri dish with a 15 mm diameter glass O-ring (for brain slices adhered to a net)

20 mL glass scintillation vial (for brain slices adhered to a net)

1.5 mL centrifuge tubes and tube rack (for brain slices not adhered to a net)

2.2 Agar Embedding and Vibraslicing

The thickness of most acute brain slices is between 300 and 500 μm. To increase the effectiveness of post-fixation, dehydration, and infiltration, embed the fixed tissue in agarose and slice into 70–80 μm sections on a vibratome. The agarose will stabilize the tissue and also keep it flat during later processing steps.

1. *Reagents*: 5 % agarose in 0.1 M phosphate buffer (20 mL total)

10 mL 0.2 M phosphate buffer (pH 7.4)

1 g agarose

~10 mL ultrapure water (18.2 MΩ · cm) (q.s. to final volume of 20 mL)

2. *Equipment* (Fig. 1a):

Stirring hot plate

Small stir bar

50 mL beaker

20 mL glass scintillation vial

Plastic transfer pipettes

Fine-tipped paint brush

Slide with two broken slides glued to each end (depth of 2 mm)

Slide with two coverslips glued to each end (depth of <400 μm)

Two or three plain glass slides

Crazy/super glue

24-well tissue culture plate

Vibratome (with tools for fixed tissue)

2.3 Processing

Processing tissue for electron microscopy involves a few key stages including post-fixation with osmium tetroxide, dehydration with ethanol or acetone, and infiltration and embedding with an epoxy resin. For optimal preservation and contrast of membranes for ssEM we have included a reduced osmium step (osmium

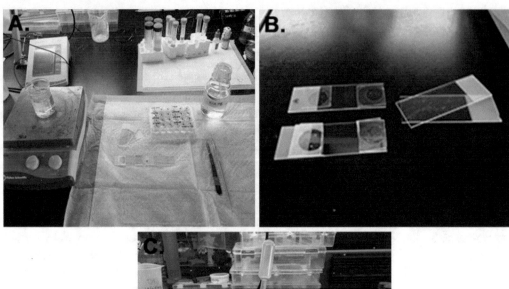

Fig. 1 Agar embedding. (**a**) Setup for agar embedding requires a 5–7 % solution of agarose maintained at 60 °C, a labeled 24-well plate with filled with buffer, a scalpel or razor blade, a paint brush, and slides for embedding. (**b**) Slides for agarose embedding can be made to accommodate different thicknesses of tissue. Coverslips work well for embedding 70 μm thick sections of tissue prior to processing while broken pieces of glass slides are useful for embedding larger pieces of tissue prior to vibratoming. (**c**) Keep the agarose in liquid form by maintaining it in a water bath at 60 °C. Cut the tip off of the plastic transfer pipette to ensure that it does not get clogged

tetroxide + potassium ferrocyanide) and added a 1 % uranyl acetate (UA) solution to the dehydration steps. All of these chemicals are extremely toxic and must be handled in a fume hood while wearing gloves.

1. *Reagents*: For processing ~10–12 vials of tissue

   0.1 M Sodium cacodylate buffer

   Ultrapure water (18.2 MΩ · cm)

   100 % Ethanol (EM grade)

   Propylene oxide (PO)

   EMbed 812 (EMbed 812, NMA, DDSA, DMP-30)

2 % Uranyl acetate (UA) in ethanol (1 g UA in 50 mL 100 % ethanol)

8 mL Reduced osmium tetroxide (1.5 % Potassium Ferrocyanide (KFeCn) + 1 % Osmium Tetroxide (OsO_4)) in 0.1 M sodium cacodylate buffer (2 mL 4 % OsO_4 + 4.8 mL buffer + 1.2 mL 10 % KFeCn)—Prepare immediately before use.

8 mL 1 % Osmium Tetroxide in 0.1 M sodium cacodylate buffer (2 mL 4 % OsO_4 + 6 mL buffer)

6 mL 50 % Ethanol (3 mL 100 % ethanol + 3 mL ultrapure water)

6 mL 50 % Ethanol + 1 % UA (3 mL ultrapure water + 3 mL 2 % UA). Filter immediately before use.

6 mL 70 % Ethanol + 1 % UA (1.8 mL ultrapure water + 1.2 mL ethanol + 3 mL 2 % UA). Filter immediately before use.

6 mL 90 % Ethanol + 1 % UA (0.6 mL ultrapure water + 2.4 mL ethanol + 3 mL 2 % UA). Filter immediately before use.

6 mL 100 % Ethanol + 1 % UA (3 mL ethanol + 3 mL 2 % UA). Filter immediately before use.

6 mL 1:1 EtOH:PO (3 mL ethanol + 3 mL PO)

9 mL 1:2 EtOH:PO (3 mL ethanol + 6 mL PO)

6 mL 1:1 PO:Epon (3 mL PO + 3 mL EMbed 812 + DMP-30)

9 mL 1:2 PO:Epon (3 mL PO + 6 mL EMbed 812 + DMP-30)

2. *Equipment*:

5–6 150 mL Tri-Pour beakers

18 20 mL Scintillation vials

2 50 mL Falcon tubes

~8 10 mL glass pipettes

Pipettor

4 20 mL Syringes

4 Disc syringe filter tips (0.2 μm pore size)

6–12 2 mL vials or 1.5 mL centrifuge tubes

Embedding molds

Stir plate

Shaker or rotator

60 °C Oven

2.4   Sectioning

Cutting and collecting serial sections requires patience and practice, but the amount of data that can be gleaned from a series will make it worthwhile. The goal is to have uniform section thickness along fold free ribbons and to pick them up on film-coated slot grids without losing any sections. Formvar and other film-coated slot grids are commercially available. However, it is less expensive in the long run to purchase uncoated slot grids and coat them yourself. For an excellent tutorial on this process as well as videos on cutting and collecting serial sections, please see the Supplemental Materials of Harris et al., [36].

1. *Reagents*:

   Filtered ultrapure water (18.2 MΩ · cm)

   1 % Toluidine Blue (1 part 70 % ethanol to 1 part 1 % Toluidine Blue)

2. *Equipment*:

   Ultra-Microtome surrounded by a plastic/plexiglass shield

   35° diamond knife—minimizes section compression

   Coated slot grids—2 mm × 1 mm slot grids coated either with formvar or pioloform

   Eyelash

   Curved fine-tip forceps—Numbered so that grids will be kept in order

   Rubber o-rings

   Grid box

   10 mL Syringe with a filter tip

   Razor blades

   Glass slides

2.5 Staining

Once sections are dry, staining with uranyl acetate and lead citrate can be done to enhance the contrast of membranes and organelles. However, the use of reduced osmium and en block staining with uranyl acetate during dehydration may provide sufficient contrast, depending on your sample and the electron microscope.

1. *Reagents*: Saturated Aqueous Uranyl Acetate

   6.25 g UA

   100 mL Ultrapure water (18.2 MΩ · cm)

2. *Reagents*: Lead Citrate

   1.33 g Lead nitrate

   1.76 g Sodium citrate

   8 mL 1 N sodium hydroxide

   ~42 mL boiled ultrapure water (18.2 MΩ · cm) (Start off with ~30 mL then q.s. to final volume of 50 mL)

3. *Equipment*:

　　6 10 mL syringes

　　6 syringe filter tips

　　50 mm petri dish

　　Dental wax

　　Razor blade

　　Filter paper cut into small wedges

　　Fine-tip forceps

2.6 Imaging

Imaging serial sections can be done on any type of transmission electron microscope operating at a voltage between 80 and 120 kV with a digital camera system. A TEM with a rotational specimen holder helps keep all of the sections in the same orientation for imaging, but a standard specimen holder such as those found in most electron microscopy core facilities will also work because the images can be rotated after the fact. Recent work using a scanning electron microscope ran in transmission mode has opened up even more possibilities for large volume reconstructions of brain tissue due its capabilities of scanning large areas of each section (64 μm × 64 μm × 10 μm or larger) at 2 nm resolution [31].

3 Procedures

3.1 Fixation

1. Before fixing any tissue, it is important to determine the microwave settings that will yield a final temperature of 35–50 °C. Place water loads in the microwave (three 500 mL plastic bottles filled with water or a load cooler work well) and use a small petri dish with ~5 mL of water placed in the center of the microwave to determine the wattage and time duration that yields the appropriate temperature. We have found that 750 W for 10–30 s are good parameters.

2. Prepare the fixative (6 % glutaraldehyde and 2 % paraformaldehyde in 0.1 M sodium cacodylate buffer) in advance and add 5 mL to the small petri dish with the glass o-ring (for brain slices adhered to a net from an interface chamber) or ~500 μL to the centrifuge tubes in a tube rack (for brain tissue not adhered to a net).

3. Run the microwave for ~30 s with a neon bulb array in the center to make sure there are no hot spots.

4. Once the experiment is complete, quickly transfer tissue to the fixative and microwave it at the predetermined settings.

5. After the microwave is done, transfer brain slices to a 20 mL scintillation vial with the same fixative or leave tissue in centrifuge tubes.

6. If processing tissue the next day, leave it at room temperature overnight. If not processing for a few days, keep the tissue at 4 °C.

3.2 Agarose Embedding and Vibraslicing

1. Use a paint brush to gently detach the brain slice from the net.

2. Rinse tissue 3 × 10 min in 0.1 M phosphate buffer

3. Prepare 5 % Agarose in 0.1 M phosphate buffer in a 20 mL scintillation vial with a stir bar by immersing the vial in a small 50 mL beaker of water on a stirring hot plate. Heat the solution to ~100 °C while rapidly stirring to melt the agarose.

4. Once the agarose is melted, turn the temperature on the hot plate down until the solution reaches ~60 °C (Fig. 1c).

5. Transfer the tissue to the slide with the double slide spacers on either end (depth ~2 mm) (Fig. 1b) and remove any excess buffer.

6. Add a drop of the agarose to the tissue using a plastic transfer pipette with the tip cut off (to prevent clogging) and place a plain slide on top to flatten the agarose.

7. Allow the agarose to cool and harden then add a few drops of buffer to facilitate the removal of the top slide.

8. Trim the agarose around the tissue into a rectangle, flip the tissue over with a paint brush, and repeat steps 5–7.

9. Trim the agarose around the tissue into a rectangle again and then glue it to the vibratome platform. If the vibratome is also used for fresh tissue, make sure to use a boat, platform, and knife holder that are exclusively for fixed tissue. Ensure that the tissue is oriented in the proper position to yield the desired plane of section.

10. Once the glue is dry, secure the platform in the vibratome, fill the boat with enough 0.1 M phosphate buffer to cover the tissue and agarose, and position the blade so it will cut across the entire surface of the tissue.

11. Cut the tissue at 70–80 μm and use a paint brush or transfer pipette to transfer the sections to a 24 well plate filled with 0.1 M phosphate buffer.

12. Once the desired sections have been acquired, transfer those sections one at a time to the slide with the double coverslip spacers on either end (depth <400 μm) and repeat steps 5–7.

13. Trim the thin layer of agarose into a rectangle around the tissue and return it to the 24 well plate. Embedding tissue in the thin layer of agarose will keep it stable and flat during subsequent processing and will not interfere with post-fixation, dehydration, or infiltration of the epoxy resin.

3.3 Processing: All Steps Are Done in a Fume Hood!

1. Transfer tissue to 2 mL vials or 1.5 mL centrifuge tubes and rinse with 0.1 M sodium cacodylate buffer 3 × 10 min.

2. Begin mixing Epon components as per kit instructions in Tri-Pour beakers. Once DMP-30 has been added, allow Epon to mix for ~15 min before adding to 1:1 and 1:2 PO mixtures.

3. Label Tri-Pour beakers for uranyl acetate waste and osmium waste. Remember to dispose of waste properly as instructed by the institution's environmental health and safety regulations.

4. Add 0.1 M sodium cacodylate buffer to one Tri-Pour beaker and ultrapure water to another Tri-Pour beaker. These beakers of solution can be used to prepare the other reagents as well as for the rinsing steps.

5. Label 20 mL scintillation vials for all of the reagents that will be used during processing.

6. Use 10 mL glass pipettes to measure out all components of the reagents except for osmium, which typically comes in small ampules. Use a small tipped transfer pipette to add osmium to appropriate reagents. Prepare reduced osmium mixture (1 % OsO_4 + 1.5 % KFeCn) last.

7. For the uranyl acetate/ethanol dehydration steps, filter each solution using the 20 mL syringe and disc syringe filter tip (0.2 μm pore size) immediately before use.

8. All processing steps are carried out at room temperature. Use fresh plastic transfer pipettes to remove and add solutions to the tissue vials. A shaker or rotator can be used during each of the steps, but is not necessary. All steps are carried out in the hood and all waste must be disposed of according to the institution's environmental health and safety regulations.

9. Once all of the reagents are prepared, use a plastic transfer pipette to suck out the cacodylate buffer in each vial of tissue and replace it with the reduced osmium mixture (1 % OsO_4 + 1.5 % KFeCn). Leave for 30 min.

10. Rinse with 0.1 M sodium cacodylate buffer five times.

11. 1 % OsO_4 for 1 h.

12. Rinse with 0.1 M sodium cacodylate buffer 5 × 2 min.

13. Rinse with ultrapure water two times.

14. Rinse with 50 % EtOH.

15. 50 % EtOH/UA for 15 min.

16. 70 % EtOH/UA for 15 min.

17. 90 % EtOH/UA for 15 min.

18. 100 % EtOH/UA for 15 min.

19. 100 % EtOH for 15 min.

20. 1:1 EtOH:PO for 10 min.

21. 1:2 EtOH:PO for 10 min.

22. PO for 2 × 15 min.

23. 1:1 PO:Epon for 1 h.

24. 1:2 PO:Epon for 1 h or overnight.

25. Prepare fresh Epon with DMP-30 according to kit instructions.

26. 100 % Epon for 3 × 1 h.

27. Prepare labels for embedding. Either use pencil or a laser printer because ink pens will smear. Add labels to the embedding mold.

28. Add some Epon to the embedding mold and embed tissue, positioning it so that the desired plane of sectioning is perpendicular to the end of the mold.

29. Top off Epon in the molds and place in a 60 °C oven for 48 h to polymerize.

3.4 Sectioning

1. Secure the block of tissue in the ultramicrotome chuck and place chuck in the trimming stand.

2. Place trimming stand onto the stage area of the ultramicrotome and secure into place (Fig. 2a).

3. Looking at the block surface through the eyepieces of the ultramicrotome, use a razor blade to trim excess Epon away from the tissue, rotating chuck and ending with a trapezoid shape (Fig. 2b).

4. Once the block is trimmed, transfer the chuck to the specimen arm of the ultramicrotome and secure it tightly. Unless the tissue is tilted, set the angle of the block/specimen arm at 0° and rotate the block so that the long edge of the trapezoid will be parallel with the knife's edge.

5. For trimming tissue to a specific depth, use a glass knife or an older diamond knife that is nicked to dry section the block until the area of interest is reached.

6. Secure the knife onto the ultramicrotome stage and adjust its position until the tissue is over part of the knife edge. Slowly advance the knife while rocking the specimen arm of the microtome up and down and using the reflection on the block face to determine how close the knife is to the tissue. Before getting too close, set the cutting window and speed on the ultramicrotome so that the block begins and ends the slow phase of the cutting well above and below the edge of the knife. Finally, advance the knife until the reflection on the block face is almost gone and switch to the automatic cutting. For this step, the

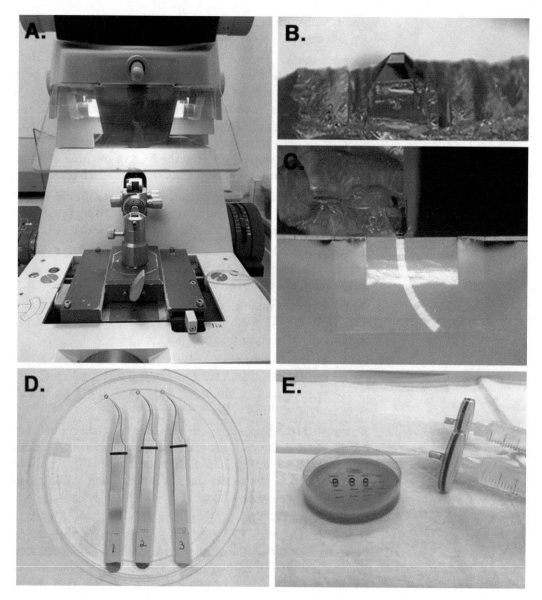

Fig. 2 Sectioning and staining. (**a**) An ultramicrotome with the chuck holding the block secured in the stand for trimming. (**b**) An epon block trimmed to form a trapezoid around the tissue. (**c**) Serial sections being cut on the diamond knife. Note the slight unevenness in section color which can be decreased by enclosing the knife and epon block in a plastic or plexiglass shield. (**d**) Three slot grids secured in curved fine-tip forceps labeled in order of section pick-up. (**e**) Setup for staining includes a small petri dish with small slits cut into pink dental wax where the slot grids can be gently placed. Make sure that the slot with the sections is above the lip of the slit. The uranyl acetate (*yellow solution*) and lead citrate (*clear solution*) should be filtered before applied in drops to the front of the grids containing the sections

specimen can be cut quickly (50 mm/s) until the appropriate depth is reached.

7. For semi-thin sections (~1 µm), use a glass knife with a boat, a histo knife, or an older diamond knife. Using a syringe with a filter tip, fill the boat with ultrapure water until it forms a convex meniscus. Use the eyelash tool to swipe along the knife's edge and help adhere water to the edge of the blade. Then use either a fresh syringe or the provided water siphon to slowly lower the water level from the back of the boat until the surface of the water takes on a smooth silvery surface.

8. Repeat step 6 but slow the cutting speed down to ~1 mm/s before switching to automatic cutting. Stop cutting once a few sections have been cut and are floating on the water.

9. Add a drop of water to a glass slide and use the eyelash tool or a loop tool to transfer the sections from the boat and to the drop on the slide. Place slide on a hot plate and allow sections to dry.

10. Add a drop of toluidine blue to the sections, return to the hot plate and wait ~1 min. Rinse dye off the slide and check the sections on an upright microscope.

11. For ultrathin sectioning, reset the ultramicrotome for 50–65 nm section thickness. Use a sharp diamond knife that is reserved for cutting ultrathins and repeat step 7 to fill and adjust the water level in the boat.

12. Repeat step 6 and once the sections are cutting, set the counter on the microtome to track the number of sections. Sections should have an even gray/silver appearance (Fig. 2c). For uniform section thickness, shield the knife and cutting arm with a plastic/plexiglass box.

13. Once a ribbon of 100–200 serial sections has been cut, clean the eyelash tool with 70 % ethanol and a Kimwipes, then gently poke at the end of the ribbon to break off 10–20 sections (depending on the size of the sections). Use the eyelash to corral the small ribbon of sections to a different part of the boat away from the main ribbon.

14. Use fine-tip forceps to grasp a formvar or pioloform coated slot grid, being careful not to poke a hole in the film, and secure the grid by sliding the rubber o-ring up the shaft of the forceps until tips are held closed.

15. Insert the edge of the grid into the boat at an angle perpendicular to the water's surface and submerge the grid to just cover the slot. Maneuver the grid near the ribbon being picked up and gently wave the grid back and forth to draw the sections closer. Once the sections are positioned over the slot, scoop the grid and sections up out of the water and confirm under the

binoculars of the ultramicrotome that the sections are in the middle of the slot.

16. Place the forceps with the grid and sections in a clean, dry petri dish and cover with the lid to allow the sections to dry without collecting any dust (Fig. 2d).

17. Repeat steps 12–15 until all of the sections are collected. If there are more grids than forceps available to hold them, the dry grids can be placed in the slots of the staining dish (see below) in order!

3.5 Staining

1. To prepare a staining dish, melt a small amount of pink dental wax and pour it into a 50 mm petri dish, to a depth of ~5 mm.

2. Once the wax has cooled, use the edge of a clean razor blade to cut slits into the wax that are ~5 mm long and ~1 mm deep.

3. For the uranyl acetate and lead citrate stains, aliquot them into 10 mL syringes with filtered tips and keep them at 4 °C once they have been prepared.

4. Gently slide the o-ring down the forceps holding the first grid of the series, but keep the forceps pinched shut. Place the grid in the first slot so that the grid is perpendicular to the surface of the wax and the slot is above the edge of the slit (Fig. 2e). Gently tap the grid to make sure it is in place and will not fall out.

5. Repeat step 4 with all of the grids until all of the slots in the staining dish are filled.

6. Add a drop of uranyl acetate to the side of the grid with the sections. Place the lid on the petri dish and wait for 5 min.

7. Rinse the uranyl acetate off the grids by holding the staining dish at an angle and gently running a stream of ultrapure water across the grids using a squeeze bottle for ~30 s. Make sure the run-off from the staining dish goes into an appropriate waste container.

8. Use small wedges of filter paper to delicately wick away any remaining water from the grids by holding the paper at the edge of the grid (don't touch the slot of the grid!).

9. Place a small piece of filter paper in the lid of the staining dish and saturate it with 1 N NaOH.

10. Add a drop of lead citrate to the side of the grid with the sections. Place the lid on the petri dish and wait for 5 min.

11. Rinse the lead citrate off by first running a few drops of 1 N NaOH across the grids and then following up with a gentle stream of ultrapure water for ~30 s. Make sure the run-off from the staining dish goes into an appropriate waste container.

12. Repeat step 8 until the grids are completely dry. Depending on when the grids are to be imaged, they can be left in the staining dish. It helps to minimize their handling to prevent tearing or damaging the film.

3.6 Imaging

1. Image a calibration grid at the magnification at which the series will be imaged. A magnification around $5000\times$ works well for capturing a relatively large field with enough resolution to unequivocally identifying synapses and subcellular structures.

2. Check all of the grids before starting to image the series to make sure that all of the grids are intact and in order (see notes for more details).

3. Begin with the first section on the first grid. Locate the area of interest for imaging. Identify fiduciary markers (mitochondria, cell bodies, etc.) to help locate the same area on subsequent sections. Name images in a sequential manner.

4. Find the fiduciary markers on each section, updating them as the series progresses, and image the same area of interest.

4 Typical/Anticipated Results

Once the images have been acquired, a variety of software can be used for subsequent alignment and analysis, including IMOD (http://bio3d.colorado.edu/imod/), FIJI (http://fiji.sc/Fiji), and Reconstruct™ (available for free download at http://synapses.clm.utexas.edu/tools). It is important that everything is scaled and calibrated properly for accurate quantifications. In this section some examples will be provided of the structures that can be identified and analyzed through serial sections. All of the examples were analyzed and reconstructed using Reconstruct™.

4.1 Dendritic Spines

The size and shape of dendritic spines can vary greatly within the same region (Fig. 3a). Tracking and quantifying the dimensions of these structures through serial sections can provide both basic anatomical data (Fig. 3f) and a measure of structural plasticity. The majority of spines have a constricted neck and either a large, mushroom-shaped head (>0.6 μm in diameter, Fig. 3b, f) [37–40] or a smaller, thin head (Fig. 3c, f). Other types include stubby spines that lack a constricted neck (Fig. 3d, f), branched spines with two or more protrusions that can have distinct morphologies (Fig. 3e), or multi-synaptic spines with multiple synaptic contacts along the head and neck. These categories can have functional implications and can only be accurately identified through serial sections.

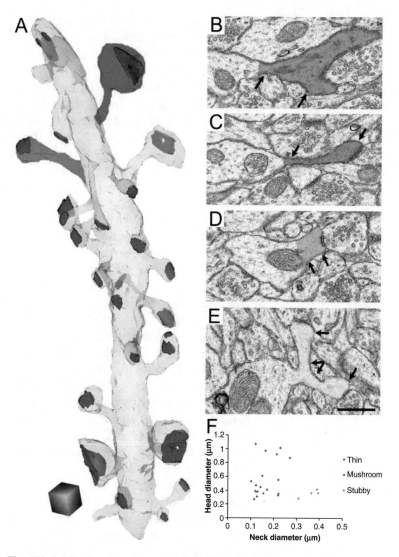

Fig. 3 Variability in spine shape and size. (**a**) three-dimensional reconstruction of a hippocampal dendrite (*gray*) illustrating different spine shapes including mushroom (*blue*), thin (*red*), stubby (*green*), and branched (*yellow*). PSDs (*red*) also vary in size and shape. Scale cube = 0.125 μm³. (**b**) An example of a mushroom spine (*blue*) with a head diameter exceeding 0.6 μm and a narrow neck. (**c**) An example of a thin spine (*red*) with a small head and narrow neck. (**d**) An example of a stubby spine (*green*) with an equal head and neck diameter and an overall length that equals its width. (**e**) An example of a branched spine (*yellow*) where both branches are thin spines. Scale bar = 0.5 μm and *arrows* indicate where the head and neck diameters were measured for each spine in (**b**)–(**e**). (**f**) A graph plotting the ratio of head diameters to neck diameters for the spines on the dendrite reconstructed in (**a**). Mushroom spines (*blue diamonds*), stubby spines (*green diamonds*), and thin spines (*red diamonds*) segregated into distinct groups. Both branches of the branched spine were of a thin shape and are situated among the thin spine dimensions (*yellow diamonds*). Adapted from Bourne and Harris, 2008 with permission from Annual Reviews

4.2 Synapses

Synapse size and shape are ultrastructural features that have a strong correlation with synaptic strength [8–13, 15, 16, 18, 24, 25, 41, 42]. The two primary types of synapses are excitatory and inhibitory that have distinctive morphological properties (Fig. 4Ai–Aiv). Excitatory synapses are characterized by a fuzzy, electron-dense structure called a postsynaptic density (PSD) that extends beneath the postsynaptic membrane and is apposed to round, clear synaptic vesicles docked at the presynaptic active zone. The PSD is usually thicker than the active zone resulting in classification as an "asymmetric" synapse, or Gray's Type 1 synapse (Fig. 4Ai–Aiv, black arrow). In contrast, inhibitory synapses have a thinner PSD apposed to a presynaptic active zone and vesicles with a more flattened, pleiomorphic appearance. The similar dimensions of the PSD and active zone led to the categorization as "symmetric" synapses, or Gray's Type 2 synapses (Fig. 4Ai–Aiv, white arrow). However, there can be a broad spectrum of presynaptic and postsynaptic thickening in the same brain region [43], further demonstrating the usefulness of analyzing synapses with ssEM. Another aspect of synaptic morphology that can be quantified with ssEM is whether the synapse has a continuous, macular surface or has a noncontinuous, perforated appearance (Fig. 4Bi–Biv), which may have functional implications regarding synapse strength and receptor composition [41, 42]. For branched spines, ssEM can also be used to determine whether the branches synapses on the same or different presynaptic axons (Fig. 4Ci–Civ). In addition to synapses, other postsynaptic specializations such as cell adhesion junctions [44] and nascent zones, areas of the PSD that lack apposing presynaptic vesicles, can be identified [13].

On the presynaptic side, ssEM is useful for determining the number of postsynaptic partners with which axonal boutons form contact (Fig. 5a–c). Within individual boutons, docked vesicles, where the vesicle membrane is in contact with the presynaptic membrane, and non-docked vesicles can be counted (Fig. 5d, e). The docked vesicle pool is thought to represent the "readily releasable" pool of vesicles and provides an ultrastructural measure of the probability of release at a synapse. The non-docked vesicles form part of the "reserve" pool that may be recruited under different physiological conditions [12, 45–49]. ssEM also increases the probability of identifying and quantifying relatively sparse presynaptic structures such as dense core vesicles that may be functionally important for expanding the active zone [12, 50].

4.3 Organelles

Local protein synthesis is required for many types of synaptic plasticity and is evidenced ultrastructurally by polyribosomes [9–11]. Polyribosomes consist of a group of three or more ribosomes and can occur in various configurations including clusters, spirals, and staggered lines. Ribosomes have opaque centers that are ~10–25 nm in diameter and surrounded by a fuzzy gray halo (Fig. 6Aii, Aiii, Bii). Individual ribosomes typically do not occur

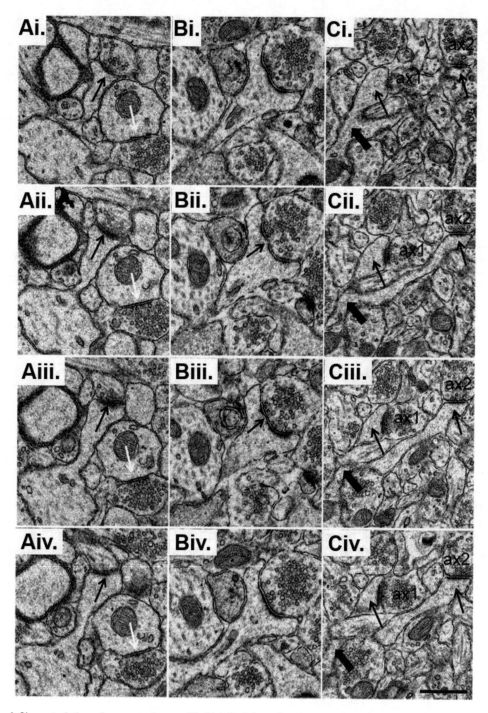

Fig. 4 Characteristics of synapse types. (**Ai–Aiv**) Serial section images through most of the macular PSD (*black arrow*) of a thin spine and through most of a symmetric synapse (*white arrow*). (**Bi–Biv**) Serial section images through some of a mushroom spine illustrating a small perforation in the PSD (*black arrow*). (**Ci–Civ**) Serial sections through the branch point (*thick arrows*) of a spine. Two thin spine heads (*thin arrows*) that form macular synapses on two different presynaptic axons (ax1 and ax2). Scale bar = 0.5 μm. Reproduced from Bourne and Harris, 2011 with permission from Wiley-Liss, Inc.

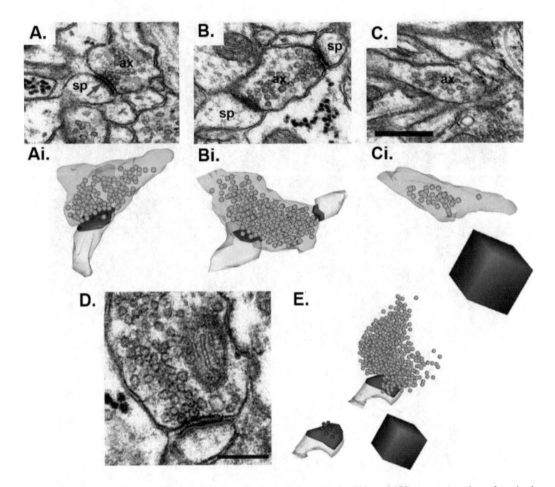

Fig. 5 3D analyses of axonal boutons and synaptic vesicles. (**a**) An EM and (**Ai**) reconstruction of a single synaptic bouton (ax) filled with vesicles (*green arrows*) synapsing with the PSD (*red arrow*) on a dendritic spine (sp). (**b**) An EM and (**Bi**) reconstruction of a multisynaptic bouton (MSB) and both postsynaptic spine partners. (**c**) An EM and (**Ci**) reconstruction of a nonsynaptic bouton (NSB). Scale bar = 0.5 μm. Scale cube = 0.125 μm^3. (**d**) EM and (**e**) 3D reconstruction of a dendritic spine (sp) and presynaptic bouton (ax) with the docked vesicles stamped in *blue*, the vesicle pool stamped in *green*, and PSD outlined in *red*. Scale bar = 0.25 μm. Scale cube = 0.0156 μm^3. Adapted from Bourne et al., 2013 with permission from Wiley Periodicals, Inc.

in more than one section because of their small size. Polyribosomes occur every 1–2 μm along the length of immature and mature hippocampal dendrites and can be found both in the shaft of the dendrite (Fig. 6Ai–Aiv) or in the head or neck of dendritic spines (Fig. 6Bi–Biv).

Smooth endoplasmic reticulum (SER) is a complex membranous network within neurons involved in intracellular calcium regulation and trafficking of membrane-bound cargo. ssEM allows the quantification and characterization of the complexity of SER within the dendritic shaft (Fig. 7a) [28] and the frequency with which SER enters dendritic spines [27]. In the dendritic shaft, SER can appear as small circles of membrane on single sections (Fig. 7a, top left)

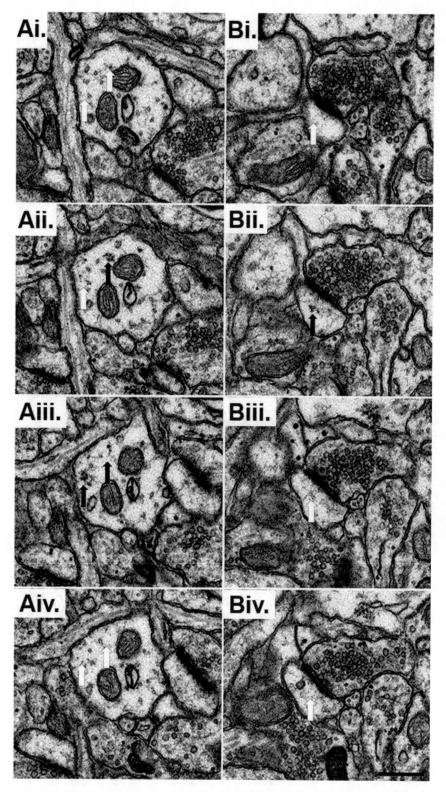

Fig. 6 Identification of polyribosomes by viewing through adjacent serial thin sections of (**Ai–Aiv**) a dendritic shaft and (**Bi–Biv**) a dendritic spine. Two polyribosomes were identified in this segment of the dendritic shaft

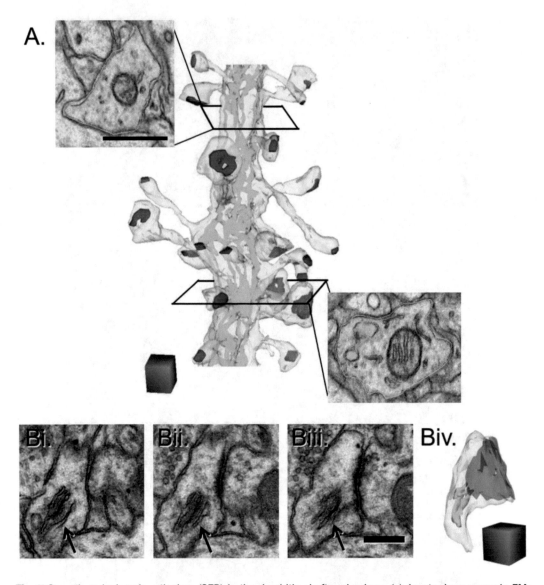

Fig. 7 Smooth endoplasmic reticulum (SER) in the dendritic shaft and spines. (**a**) *Insets* show example EMs from the reconstructed dendrites (*yellow*) with simple and complex tubules of ER (*green*). Scale bar = 0.5 μm. Scale cube = 500 nm on a side. (**b**) Spine apparatus in a mushroom spine. (**Bi–Biii**) Serial sections through a mushroom spine containing a spine apparatus (*arrows*). Scale bar = 0.5 μm. (**Biv**) Reconstruction of the same spine (*yellow*) with its PSD (*red*) and spine apparatus (*green*). Scale cube = 0.125 μm^3. Adapted from Cui-Wang et al., 2012 with permission from Elsevier Inc.

Fig. 6 (Continued) (*Black arrows* on sections **Aii** and **Aiii**). One polyribosome was identified in the head of the dendritic spine (*Black arrow* on section **Bii**). *White arrows* indicate the location of PR on adjacent sections where no structures are present that resemble PR, indicating their existence in only 1 or 2 sections, consistent with the number of individual ribosomes in each cluster. Scale bar = 0.5 μm. Reproduced from Bourne and Harris, 2011 with permission from Wiley-Liss, Inc.

that connect across serial sections to form tubules. SER can also appear as larger, more amorphous blobs that form pools of membrane (Fig. 7a, lower right). ssEM is often necessary to distinguish SER from endosomes, structures that can look very similar on single sections [51]. In areas of the brain where SER infrequently enters dendritic spines, such as in stratum radiatum of area CA1 of the hippocampus where fewer than 20 % of spines contain SER [27], ssEM permits the quantification and characterization of these relatively rare occurrences. SER in spines can either be in the form of a simple tubule or in the more complex form of a spine apparatus that contains stacks of membrane separated by dense plate material (Fig. 7Bi–Biv).

5 Troubleshooting/Notes

5.1 Fixation

1. Leaving the glutamate out at room temperature for 3 days prior to preparation of the fixative can decrease the occurrence of "pepper" that appears as black specks on the final stained sections.

2. Quickly transferring the tissue from the experimental chamber to the fixative can help avoid ultrastructural artifacts caused by hypoxia of the tissue.

3. If a microwave is not available, immersion fixation can also yield good ultrastructure. It helps if the brain slices are on the thinner side (<400 μm) and if they are left at room temperature in the fixative overnight. The next day the tissue may be processed or placed at 4 °C for long term storage.

5.2 Agarose Embedding and Vibraslicing

1. After the agarose has dissolved in the buffer, make sure the temperature is ~60 °C before dropping the agarose onto the tissue. Agarose that is too hot can damage the tissue.

2. Depending on the size of the tissue, spacers of different thicknesses can be made to accommodate the agarose embedding.

3. When removing the top slide after the agarose has hardened around the tissue, adding a little buffer can prevent the accidental tearing of the tissue.

4. During processing, the osmium in the tissue will sometimes leech out into the surrounding agar, making it difficult to see the tissue. Making the agar fresh each time you do agar embedding (as opposed to reheating a previously made batch) helps to minimize this issue.

5.3 Processing

1. When preparing the epoxy resin, the bottles for the NMA and DDSA often get stuck closed. Placing the bottles in a 60 °C oven for a few minutes warms up the chemicals enough to loosen the lids.

2. It is not essential to have Tri-Pour beakers for waste if it is easier to directly pour it into the appropriate containers. However, having a wider spout to initially pour the waste into can minimize spills.

3. For the solid waste including the 10 mL glass pipettes, 20 mL scintillation vials, and plastic transfer pipettes, a large Ziploc bag kept in the fume hood during processing is a convenient way to consolidate all of the contaminated materials.

4. Use of reduced osmium tetroxide (osmium tetroxide + potassium ferrocyanide) can significantly enhance the contrast of membranes over osmium alone and is particularly useful for serial section reconstructions.

5. Many of these steps can be carried out using a specialized microwave (e.g., PELCO/Biowave/Ted Pella) to increase and speed up the penetration of reagents (see http://synapses.clm.utexas.edu for protocol). However, while processing without a microwave takes a bit longer, it yields tissue samples that are of the same high quality.

6. Thickness of the tissue sample can have a profound impact on the effectiveness of post-fixation with osmium and dehydration with ethanol. Tissue more than 1 mm in thickness often has poor secondary fixation and infiltration of Epon resulting in bad ultrastructure and sections that do not cut well. At minimum, trimming the tissue as small as possible can help, but vibraslicing the fixed tissue to a thickness <100 μm is the best way to ensure quality processing for EM.

7. When exchanging reagents, be sure to thoroughly remove the previous solution. This is particularly important for dehydration steps to completely get rid of all water from the specimen.

5.4 Serial Sectioning

1. When trimming, if the agarose is too dark to easily see the tissue, periodically take the block out of the chuck and check it under a microscope to be sure the region of interest remains intact.

2. If the tissue breaks off while trimming the trapezoid around the tissue, it can be superglued back onto the block. However, sometimes this will compromise the sectioning, making the block "stutter" as it goes across the knife. The piece of tissue can also be re-embedded into a new block with fresh Epon.

3. Trimming the trapezoid with three straight edges and one slanted edge makes it easier to orient yourself when looking at the section on the electron microscope and can help when trying to break apart the ribbon (Fig. 8a).

4. When selecting a diamond knife, one with a 35° angle tends to compress the tissue less than one with a 45° angle (Fig. 8b, c).

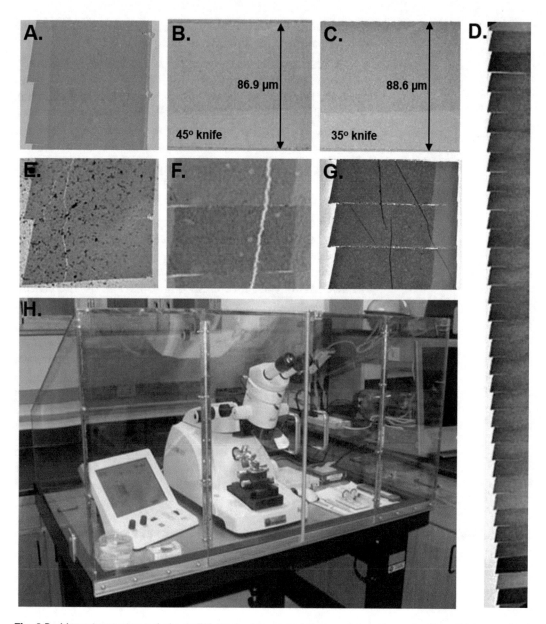

Fig. 8 Problems to overcome during ssEM. (**a**) Goal is a set of clean serial sections of uniform thickness. (**b, c**) Section compression is minimized by using a 35° diamond knife. (**d**) Uneven section thickness. (**e**) Dirty grid stain. (**f**) Poorly infiltrated tissue has cracks. (**g**) Folds in the Pioloform film. (**h**) Enclosure for the ultramicrotome usually solves most issues that cause uneven section thickness. Reproduced from Harris et al., 2006 with permission from Society for Neuroscience

5. Make sure to use filtered water in the boat of the diamond knife and keep the eyelash tool clean with 70 % ethanol to prevent it from introducing dirt and oil into the water. If more water needs to be added to the boat, use the filtered syringe, not the siphon used to adjust the water levels.

6. If water ends up on the block face, gently dab the block face with a small piece of filter paper to dry it.

7. If serial sections are not hanging together, first try re-trimming the block. Smooth, straight edges will facilitate sections sticking together. Enclose the microtome with a plastic/plexiglass box to minimize air flow that could make the sections not stick (Fig. 8h). In some instances, a spritz of hairspray onto the face of the block will help make the sections stickier.

8. Enclosing the knife and chuck in a plastic/plexiglass box will also help to ensure that the serial sections are of an even thickness (light silver/gray) (Fig. 8d, h).

9. If the ribbon of serial sections break apart during cutting, which tends to happen if the ribbon curves and runs into the wall of the boat, the sections can still be collected and their order sorted out using the electron microscope. However, in this instance it is of even more importance that all of the sections are collected.

10. When breaking apart the ribbon, tap the sections at the seam where they are joined with the eyelash. If the sections are sticking to the eyelash, dip it in 70 % ethanol and wipe it with a Kimwipes between taps.

11. The number of sections per ribbon collected on each grid will depend on the size of the sections. If sections end up above or below the slot, break the ribbon into smaller pieces.

12. Leave the grids pinched in the forceps until they are dry and then transfer to the staining dish. Minimizing the number of times the grids are handled will decrease the likelihood of poking a hole in the film.

5.5 Staining

1. Make sure the staining dish is clean to minimize dirt and debris from ending up on the sections.

2. When adding the drops of lead citrate to the grids, try not to exhale directly on the grids. The carbon dioxide can cause the lead to precipitate.

3. Thoroughly rinse the sections after the UA and lead citrate to prevent precipitate from forming on the sections.

4. Use the filter paper to completely dry the grids after the rinses to prevent residual stain from pooling around the grids.

5. If there is a lot of lead precipitate (Fig. 8e), the grids can be soaked in double distilled water for an hour and that will help reduce the precipitate. Grids can be soaked for longer periods if the precipitate is really bad. In some cases they will need to be re-stained, preferably with freshly made UA and lead citrate.

6. Tissue that has been labeled with DAB or other electron dense precipitates can be viewed on the electron microscope without UA or lead citrate staining. This can help identify where the labeled processes are without the potential confound of lead precipitate.

7. Depending on the tissue, the combination of reduced osmium, regular osmium, and en bloc uranyl acetate may provide sufficient contrast, ameliorating the need for any post-staining.

5.6 Imaging

1. Doing a quick section check of all of the grids can prevent too much time from being wasted on a series where one of the middle grids has a hole or something that would prevent the series from being completely imaged.

2. If it is suspected that the serial sections are out of order, imaging the top corner (left or right) of the first and last section on each grid as sections are checked for holes or precipitate can help put the grids in the correct order before imaging the series.

3. Aim to image near the middle of the sections to decrease the probability of the region of interest migrating off of the section before the series is completely imaged. Try to avoid imaging areas where there are either cracks in the tissue arising from poor infiltration of epon (Fig. 8f) or folds in film coating the slot grid (Fig. 8g).

4. If sections tend to drift during imaging, switch the scope to a lower magnification and allow the beam to "bake" the section for a few minutes. This can help stabilize the film and decrease drifting.

5.7 Analysis

1. For any structure of interest (polyribosome, SER, etc.), identify for each series a best example of the organelle/structure to which others can be compared.

2. Adjusting the contrast of the images can help to identify the boundaries or connections of structures like synapses and SER.

6 Conclusions

Serial section electron microscopy is a powerful tool to examine synaptic connectivity and the distribution of organelles. Recent advances in computing technology are likely to facilitate the speed with which these types of analyses can be done. However, so far, humans are still much better at identifying the boundaries of cellular membranes and identifying ambiguous structures. Although the initial collection of these types of data is time consuming, once the data are acquired they can be revisited time and again for further analysis.

References

1. Palay SL, Palade GE (1955) The fine structure of neurons. J Biophys Biochem Cytol 1:69–88

2. Gray EG (1959) Axo-somatic and axo-dendritic synapses of the cerebral cortex: an electron microscope study. J Anat 93:420–433

3. Kosaka T, Kosaka K (2005) Intraglomerular dendritic link connected by gap junctions and chemical synapses in the mouse main olfactory bulb: electron microscopic serial section analyses. Neuroscience 131:611–625

4. Mishchenko Y, Hu T, Spacek J, Mendenhall J, Harris KM, Chklovskii DB (2010) Ultrastructural analysis of hippocampal neuropil from the connectomics perspective. Neuron 67:1009–1020

5. Bock DD, Lee WC, Kerlin AM, Andermann ML, Hood G, Wetzel AW, Yurgenson S, Soucy ER, Kim HS, Reid RC (2011) Network anatomy and in vivo physiology of visual cortical neurons. Nature 471:177–182

6. Helmstaedter M, Briggman KL, Turaga SC, Jain V, Seung HS, Denk W (2013) Connectomic reconstruction of the inner plexiform layer in the mouse retina. Nature 500:168–174

7. Kasthuri N, Hayworth KJ, Berger DR, Schalek RL, Conchello JA, Knowles-Barley S, Lee D, Vazquez-Reina A, Kaynig V, Jones TR, Roberts M, Morgan JL, Tapia JC, Seung HS, Roncal WG, Vogelstein JT, Burns R, Sussman DL, Priebe CE, Pfister H, Lichtman JW (2015) Saturated reconstruction of a volume of neocortex. Cell 162:648–661

8. Bourne JN, Harris KM (2012) Nanoscale analysis of structural synaptic plasticity. Curr Opin Neurobiol 22:372–382

9. Ostroff LE, Fiala JC, Allwardt B, Harris KM (2002) Polyribosomes redistribute from dendritic shafts into spines with enlarged synapses during LTP in developing rat hippocampal slices. Neuron 35:535–545

10. Bourne JN, Sorra KE, Hurlburt J, Harris KM (2007) Polyribosomes are increased in spines of CA1 dendrites 2h after the induction of LTP in mature rat hippocampal slices. Hippocampus 17:1–4

11. Bourne JN, Harris KM (2011) Coordination of size and number of excitatory and inhibitory synapses results in a balanced structural plasticity along mature hippocampal CA1 dendrites during LTP. Hippocampus 21:354–373

12. Bourne JN, Chirillo MA, Harris KM (2013) Presynaptic ultrastructural plasticity along CA3 → CA1 axons during LTP in mature hippocampus. J Comp Neurol 521:3898–3912

13. Bell ME, Bourne JN, Chirillo MA, Mendenhall JM, Kuwajima M, Harris KM (2014) Dynamics of nascent and active zone ultrastructure as synapses enlarge during long-term potentiation in mature hippocampus. J Comp Neurol 522:3861–3884

14. Fiala JC, Allwardt B, Harris KM (2002) Dendritic spines do not split during hippocampal LTP or maturation. Nat Neurosci 5:297–298

15. Popov VI, Davies HA, Rogachevsky VV, Patrushev IV, Errington ML, Gabbott PL, Bliss TV, Stewart MG (2004) Remodeling of synaptic morphology but unchanged synaptic density during late phase long-term potentiation (LTP): a serial section electron micrograph study in the dentate gyrus in the anaesthetized rat. Neuroscience 128:251–262

16. Stewart MG, Medvedev NI, Popov VI, Schoepfer R, Davies HA, Murphy K, Dallerac GM, Kraev IV, Rodriguez JJ (2005) Chemically induced long-term potentiation increases the number of perforated and complex postsynaptic densities but does not alter dendritic spine volume in CA1 of adult mouse hippocampal slices. Eur J Neurosci 21:3368–3378

17. Park M, Salgado JM, Ostroff L, Helton TD, Robinson CG, Harris KM, Ehlers MD (2006) Plasticity-induced growth of dendritic spines by exocytic trafficking from recycling endosomes. Neuron 52:817–830

18. Sorra KE, Harris KM (1998) Stability in synapse number and size at 2 hr after long-term potentiation in hippocampal area CA1. J Neurosci 18:658–671

19. Witcher MR, Kirov SA, Harris KM (2007) Plasticity of perisynaptic astroglia during synaptogenesis in the mature rat hippocampus. Glia 55:13–23

20. Holtmaat A, Wilbrecht L, Knott GW, Welker E, Svoboda KA (2006) Experience-dependent and cell-type-specific spine growth in the neocortex. Nature 441:979–983

21. Knott GW, Holtmaat A, Wilbrecht L, Welker E, Svoboda KA (2006) Spine growth precedes synapse formation in the adult neocortex in vivo. Nat Neurosci 9:1117–1124

22. Trachtenberg JT, Chen BE, Knott GW, Feng G, Sanes JR, Welker E, Svoboda K (2002) Long-term in vivo imaging of experience-dependent synaptic plasticity in adult cortex. Nature 420:788–794

23. Zito K, Knott G, Shepherd GM, Shenolikar S, Svoboda K (2004) Induction of spine growth and synapse formation by regulation of the spine actin cytoskeleton. Neuron 44:321–334

24. Ostroff LE, Cain CK, Jindal N, Dar N, Ledoux JE (2012) Stability of presynaptic vesicle pools and changes in synapse morphology in the

amygdala following fear learning in adult rats. J Comp Neurol 520:295–314

25. Ostroff LE, Cain CK, Bedont J, Monfils MH, Ledoux JE (2010) Fear and safety learning differentially affect synapse size and dendritic translation in the lateral amygdala. Proc Natl Acad Sci U S A 107:9418–9423

26. Harris KM, Spacek J, Bell ME, Parker PH, Lindsey LF, Baden AD, Vogelstein JT, Burns R (2015) A resource from 3D electron microscopy of hippocampal neuropil for user training and tool development. Sci Data 2, Article number 150046. Doi: 10.1038/sdata.2015.46

27. Spacek J, Harris KM (1997) Three-dimensional organization of smooth endoplasmic reticulum in hippocampal CA1 dendrites and dendritic spines of the immature and mature rat. J Neurosci 17:190–203

28. Cui-Wang T, Hanus C, Cui T, Helton T, Bourne J, Watson D, Harris KM, Ehlers MD (2012) Local zones of endoplasmic reticulum complexity confine cargo in neuronal dendrites. Cell 148:309–321

29. Briggman KL, Bock DD (2012) Volume electron microscopy for neuronal reconstruction. Curr Opin Neurobiol 22:151–161

30. Kleinfeld D, Bharioke A, Blinder P, Bock DD, Briggman KL, Chklovskii DB, Denk W, Helmstaedter M, Kaufhold JP, Lee WC, Meyer HS, Micheva KD, Oberlaender M, Prohaska S, Reid RC, Smith SJ, Takemura S, Tsai PS, Sakmann B (2011) Large-scale automated histology in the pursuit of connectomes. J Neurosci 31:16125–16138

31. Kuwajima M, Mendenhall JM, Lindsey LF, Harris KM (2013) Automated transmission-mode scanning electron microscopy (tSEM) for large volume analysis at nanoscale resolution. PLoS One 8:e59573. doi:10.1371/journal.pone.0059573

32. Wanner AA, Kirschmann MA, Genoud C (2015) Challenges of microtome-based serial block-face scanning electron microscopy in neuroscience. J Microsc 259:137–142

33. Hayworth KJ, Xu CS, Lu Z, Knott GW, Fetter RD, Tapia JC, Lichtman JW, Hess HF (2015) Ultrastructurally smooth thick partitioning and volume stitching for large-scale connectomics. Nat Methods 12:319–322

34. Jain V, Seung HS, Turaga SC (2010) Machines that learn to segment images: a crucial technology for connectomics. Curr Opin Neurobiol 20:656–666

35. Jensen FE, Harris KM (1989) Preservation of neuronal ultrastructure in hippocampal slices using rapid microwave-enhanced fixation. J Neurosci Methods 29:217–230

36. Harris KM, Perry E, Bourne J, Feinberg M, Ostroff L, Hurlburt J (2006) Uniform serial sectioning for transmission electron microscopy. J Neurosci 26:12101–12103

37. Harris KM, Jensen FE, Tsao B (1992) Three-dimensional structure of dendritic spines and synapses in rat hippocampus (CA1) at postnatal day 15 and adult ages: implications for the maturation of synaptic physiology and long-term potentiation. J Neurosci 12:2685–2705

38. Harris KM, Stevens JK (1989) Dendritic spines of CA1 pyramidal cells in the rat hippocampus: serial electron microscopy with reference to their biophysical characteristics. J Neurosci 9:2982–2997

39. Bourne J, Harris KM (2007) Do thin spines learn to become mushroom spines that remember? Curr Opin Neurobiol 17:381–386

40. Bourne JN, Harris KM (2008) Balancing structure and function at hippocampal dendritic spines. Annu Rev Neurosci 31:47–67

41. Geinisman Y, de Toledo-Morrell L, Morrell F, Heller RE, Rossi M, Parshall RF (1993) Structural synaptic correlate of long-term potentiation: formation of axospinous synapses with multiple completely partitioned transmission zones. Hippocampus 3:435–445

42. Nicholson DA, Geinisman Y (2009) Axospinous synaptic subtype-specific differences in structure, size, ionotropic receptor expression and connectivity in apical dendritic regions of rat hippocampal CA1 pyramidal neurons. J Comp Neurol 512:399–418

43. Colonnier M (1968) Synaptic patterns on different cell types in the different laminae of the cat visual cortex. An electron microscope study. Brain Res 9:268–287

44. Spacek J, Harris KM (1998) Three-dimensional organization of cell adhesion junctions at synapses and dendritic spines in area CA1 of the rat hippocampus. J Comp Neurol 393:58–68

45. Lisman J, Harris KM (1993) Quantal analysis and synaptic anatomy – integrating two views of hippocampal plasticity. Trends Neurosci 16:141–147

46. Harris KM, Sultan P (1995) Variation in the number, location and size of synaptic vesicles provides an anatomical basis for the nonuniform probability of release at hippocampal CA1 synapses. Neuropharmacology 34:1387–1395

47. Shepherd GMG, Harris KM (1998) Three-dimensional structure and composition of CA3 → CA1 axons in rat hippocampal slices: implications for presynaptic connectivity and compartmentalization. J Neurosci 18:8300–8310

48. Schikorski T, Stevens CF (2001) Morphological correlates of functionally defined synaptic vesicle populations. Nat Neurosci 4:391–395

49. Branco T, Marra V, Staras K (2010) Examining size-strength relationships at hippocampal synapses using an ultrastructural measurement of synaptic release probability. J Struct Biol 172:203–210

50. Sorra KE, Mishra A, Kirov SA, Harris KM (2006) Dense core vesicles resemble active-zone transport vesicles and are diminished following synaptogenesis in mature hippocampal slices. Neuroscience 141:2097–2106

51. Cooney JR, Hurlburt JL, Selig DK, Harris KM, Fiala JC (2002) Endosomal compartments serve multiple hippocampal dendritic spines from a widespread rather than a local store of recycling membrane. J Neurosci 22:2215–2224

Neuromethods (2016) 115: 63–80
DOI 10.1007/7657_2015_81
© Springer Science+Business Media New York 2016
Published online: 23 February 2016

Combining Anterograde Tracing and Immunohistochemistry to Define Neuronal Synaptic Circuits

Dusica Bajic

Abstract

Connectivity among different brain regions has been studied since the original neuronal descriptions by Santiago Ramón y Cajal. Ultimately, only evidence of synapse proves actual connectivity between neurons originating from different brain regions. This report focuses on technical aspects of anterograde neuroanatomical tract tracing combined with immunohistochemical methods specific for ultrastructural analysis of neuronal contacts, synapses. Specifically, this technique combines peroxidase labeling of the anterograde tracer, biotinylated dextran amine (BDA 10 kDa) with immunoperoxidase-silver enhancement detection of a neuroactive substance present in the structures postsynaptic to tracer-labeled axon terminals. This technique is widely utilized to identify neuronal circuits as well as the neurochemical content of neurons that are implicated in incredibly complex and dynamic tasks. In this report we discuss technical steps in detail, as well as the technique's advantages and limitations.

Keywords: Anterograde tracer, Biotinylated dextran amine, BDA, Electron microscopy, Immunohistochemistry, Synapse, Tyrosine hydroxylase, Ultrastructure

Abbreviations

ABC	Avidin–biotin complex
BDA	Biotinylated dextran amine
BSA	Bovine serum albumin
DAB	Diaminobenzidine
HRP	Horseradish peroxidase
OsO_4	Osmium tetroxide
PB	Phosphate buffer
PBS	Phosphate buffer saline
TBS	Tris-buffered saline
TH	Tyrosine hydroxylase

1   Introduction

Anatomical connections among different brain regions have been studied since the first decade of the twentieth century when Santiago Ramón y Cajal proposed that neurons communicate with each other [1]. As a result, a whole new subdiscipline emerged: experimental neuroanatomical tract tracing. In fact, light microscopic analysis is a first step in identifying the specific neuronal circuitry for which numerous tract-tracing methods were implemented in the past [2]. However, the limitation of light microscopic technique lies in its inability to demonstrate the actual direct contacts (synapses) between neurons of different regions. Specifically, at a synapse, the plasma membrane of the signal-passing neuron (the *presynaptic* neuron) comes into a close apposition with the membrane of the target (*postsynaptic*) neuronal component (soma, dendrite, or axon). Therefore, in this chapter, we describe a comprehensive technique to visualize synapses between neurons comprising a specific circuit. The technique utilizes a successful and widely applied neuroanatomical anterograde tracer, biotinylated dextran amine (BDA) coupled with immunohistochemistry. In other words, this combined technique identifies neurons that project to a distant area and make synaptic contacts with specific neurochemically identified neurons of interest. Only by identifying synapses between neurons of two different regions can their direct connectivity be unequivocally confirmed.

It was Veenman et al. [3] who initially reported excellent quality of BDA (specifically, *BDA 10 kDa*) as an anterograde tracer. From the site of injection in live animals, dextran molecules are taken up by perikarya and dendrites by an unknown mechanism and transported centrifugally, along their axons towards the synaptic terminals in an anterograde fashion (Fig. 1a) [4, 5]. Its advantageous properties are several and include the following [2]: (1) tolerance to a variety of fixatives; (2) does not require antibodies for identification; and (3) does not interact with immunohistochemical treatment. As a result, this widely accepted anterograde tracer, BDA, can be efficiently combined with immunohistochemical staining of interest for double-labeled light and electron microscopic purposes. Light microscopic analysis gives valuable information about the adequacy of anterograde tracer injection, as well as the pattern of anterograde projections of neurons from the injection site. Thus, light microscopic analysis is usually considered a prerequisite to embarking on electron microscopic technique. The latter involves several steps that are each sophisticated, time-consuming, and technically challenging steps on their own. Specifically, these include (1) anterograde tracer injections into the particular brain area of interest; (2) double-labeling of BDA in combination with specific neuronal immunohistochemistry to

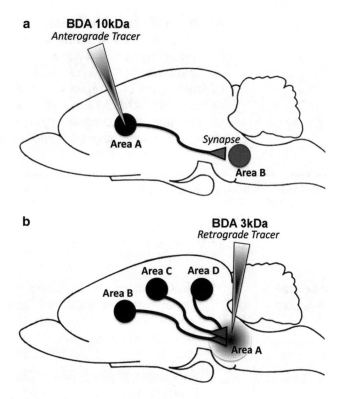

Fig. 1 Schematic representation of tract-tracing neuroanatomical studies. Biotinylated dextran amines (BDA) are organic compounds used as ANTEROGRADE (*panel* **a**) and RETROGRADE (*panel* **b**) neuroanatomical tracers. They can be used for labeling the source, as well as the point of termination of neuronal connections and therefore to study neuronal pathways. *Panel* (**a**) illustrates principles of anterograde tract tracing. A particular tracer, in this case *BDA 10 kDa*, is injected into *Area A*, picked up by the neuronal cell bodies, and finally transported down the axons all the way to its terminal field, *Area B*. Using electron microscopic technique, one can demonstrate that this anterogradely labeled neuron that originates from *Area A* forms a direct synaptic contact with another neuron located in *Area B*, thus comprising a part of the specific neurofunctional circuit. (**b**) In contrast, a retrograde tracer (e.g., *BDA 3 kDa*) from the area of injection, *Area A*, is picked up by neuronal terminals, and transported in the opposite direction from the neuronal terminal field all the way back to the neuronal cell body in a retrograde fashion (*Areas B–D*). This type of tract tracing may give information as to which areas have converging neurosynaptic input into the area of the tracer injection, *Area A*. High-molecular-weight BDA (10 kDa) yields sensitive and detailed labeling of axons and terminals (**a**), while low-molecular-weight BDA (3 kDa) yields sensitive and detailed retrograde labeling of neuronal cell bodies (**b**) [4, 5]. Therefore, BDA 10 kDa and BDA 3 kDa should not be used interchangeably

pinpoint a neurotransmitter of target neurons in a distant brain region of interest; and (3) electron microscopic analysis to finally demonstrate *direct contacts* (synapses) between anterogradely labeled axons and neurons (either soma, dendrites, or axons) in a distant brain region of interest. Schematic illustrations of the principles of double-labeling technique using BDA and immuno-histochemistry are shown in Fig. 1. In summary, we describe double-labeling technique to identify connectivity between two different anatomical regions by utilizing anterograde tracer, BDA (specifically, *BDA 10 kDa*) combined with tyrosine-hydroxylase (TH) immunohistochemistry, a marker of catecholaminergic neurons in specific regions of the brainstem.

2 Equipment, Materials, and Setup

The following section lists most of the important equipment and materials that are required for successful completion of the experiment.

2.1 Anterograde Tracer Injections

1. *Glass pipettes* (1.2 mm outer diameter, with filament; World Precision Instruments, Inc., Sarasota, FL) and a *pipette puller* to allow for creation of micropipettes with a fine tip; inner tip diameter of approximately 15–20 μm (no bigger than 40 μm to allow for discrete tracer injections in the area of interest; Fig. 1).

2. *Stereotaxic surgical frames* (Harvard Apparatus, a Harvard Bioscience company that makes many varieties for different species) for selective injections of tracer of interest in discrete anatomical regions.

3. *Anterograde tracer BDA* (*BDA 10 kDa*; 10,000 MW, D-1956; Molecular Probes, Eugene, OR). Dextrans are hydrophilic polysaccharides characterized by their moderate-to-high molecular weight, good water solubility, and low toxicity. They also generally exhibit low immunogenicity. Dextrans are biologically inert due to their uncommon poly-(α-D-1,6-glu-cose) linkages, which render them resistant to cleavage by most endogenous cellular glycosidases. Tracer should be dissolved in water, saline, or 10 mM phosphate buffer (PB), pH 7.25 (5–10 % rodents; 10 % primates).

2.2 Brain Tissue Fixative Solution

BDA is compatible with a wide range of fixatives. Considering the technique of BDA visualization is coupled with immunohistochem-istry of a particular neuronal marker, fixatives that work well for light microscopic analysis vary and may include buffered 4 % form-aldehyde, 0.1 % glutaraldehyde, and 0.25 % saturated picric acid solution [2]. However, for the purpose of ideal brain tissue fixation that will lead to optimal tissue preservation for electron microscopic analysis, we recommend the following solution to be prepared for the transcardiac perfusion (see below): *combination of 3.75 % acro-lein* (Electron Microscopy Sciences, Fort Washington, PA) and *2 % paraformaldehyde* in 0.1 M PB, pH 7.4.

2.3 Vibratome

This is an instrument similar to a microtome but uses a vibrating razor blade to cut through fixed brain tissue embedded in low-gelling-temperature agarose. For the purpose of the ultramicroscopic analysis, using a vibratome (instead of a cryostat) avoids tissue exposure to freezing temperatures and avoids dehydration and crystallization of water molecules in brain tissue prior to embedding, thus decreasing loss of cell constituents. Its other advantage is the lack of any harsh chemical treatments (e.g., paraffin embedding) that may lead to antigen instability.

2.4 Tissue Double-Labeling

1. *BDA visualization* requires the following materials: (1) Avidin–Biotin Complex (Elite Standard Vectastain ABC Kit, PK-6100, Vector Laboratories); (2) 3,3′-diaminobenzidine (DAB; Aldrich, Milwaukee, WI), and (3) 30 % hydrogen peroxide. Since the BDA label is taken up by the neurons and fills the neuronal cytoplasm, the detection reporter molecule (streptavidin-horseradish peroxidase (HRP)) must cross all cellular membranes to bind to its target, BDA. The ABC Kit is widely accepted as one of the most sensitive, economical, and reliable immunoperoxidase detection systems available. Although the visualization of BDA does not require any antibodies, the penetration of streptavidin-HRP conjugate from the ABC Kit takes some time to occur; incubation time is 1.5–2 h. The addition of any detergents (e.g. 0.3–0.5 % Triton X-100) that might improve penetration is not necessary for visualization of BDA, and should be absolutely avoided for any tissue processing performed for electron microscopic analysis.

2. *Immunohistochemical labeling* of neurons of interest must include (1) primary antisera[1] that should be diluted in 0.1 % bovine serum albumin (BSA) in 0.1 M Tris-buffered saline (TBS) solution at a certain concentration; (2) blocking solution (0.01 M phosphate-buffered saline with 0.2 % gelatin (Gelatin; IGSS quality; Amersham Life Sciences, Piscataway, NH) and 0.8 % BSA); (3) appropriate secondary antibody[2] that must be conjugated to 1 nm gold particles diluted in the same blocking solution; (4) rinsing solutions: 0.1 M TBS, 0.01 M phosphate-buffered saline (PBS; pH 7.4), and 0.2 M sodium citrate buffer solution (pH 7.4); as well as (5) silver enhancement kit (Intense M Sliver Enhancement kit, RPN 491, Amersham Corp.).

---

[1] In our previously published work [6–8], we used mouse primary antisera directed against tyrosine hydroxylase (TH; Incstar Corp., Stillwater, MN) that was diluted 1:5000 with 0.1 M TBS and contained 0.1 % BSA.

[2] The secondary antibody used in our experiments that visualized catecholamine neurons was goat anti-mouse secondary antibody conjugated to 1 nm gold particle (Auro Probe One GAM; RPN 471; Amersham Corp., Arlington Heights, IL) diluted 1:50 with the blocking solution.

2.5 Electron Microscopy

Solutions and setups necessary for different steps leading to a tissue embedding for electron microscopy include the following:

1. *2 % osmium tetroxide (OsO₄) solution in 0.1 M PB* is a staining agent that is widely used for transmission electron microscopy to provide contrast to the image [9]. It embeds a heavy metal, creating a high electron-scattering rate. In the staining of the cell membranes, osmium binds directly into cell membranes to create contrast with the neighboring cytoplasm. It also stabilizes many proteins by transforming them into gels without destroying structural features while preventing protein coagulation by alcohols during dehydration [10] (see next).

2. Series of *diluted alcohol solutions* to dehydrate the tissue: 50, 70, 95, 100 % alcohols.

3. After dehydration, the tissue needs to be embedded (hardened), so it can be appropriately sectioned prior to viewing. Both "transition solvent" such as *propylene oxide* and an epoxy resin, *Embed 812* (cat# EMS 14120; Electron Microscopy Sciences, Hatfield, PA; also previously known as Epon 812) are needed. The latter will be used for embedding/infiltration of tissue. Finally, flat embedding should be done between two *sheets of plastic* (Aclar, Ted Pella, Inc., Redding, CA) under a higher temperature; the *oven* should be set at 65 °C.

4. *Ultramicrotome* (Leica Microsystems) provides easy preparation of extremely thin (ultrathin) tissue sections of samples for electron microscopic analysis.

5. *Electron microscope.*

3 Procedures

3.1 Anterograde Tracer Injection

The first step in obtaining successful results is to identify the exact location of the brain area of interest whose projection one wants to study as a part of a circuit. For example, in the hypothesis driven experiment, one would want to show that brain *Area A* projects to and forms synapses with neurons from the brain *Area B*, located at a distance to *Area A* (Fig. 1a). How successful one is going to be in showing the projection of interest is primarily determined by the accuracy of stereotaxic coordinates for brain *Area A*, the tracer injection site.

Once the coordinates are determined and the surgical stereotaxic apparatus set, animals should be deeply anesthetized with pentobarbital (50 mg/kg). The incisor bar should be set at −2.5 mm. The dorsal surface of the rat head should be shaved and skin surgically prepared using aseptic techniques prior to using a scalpel to cut it. Once the skull is exposed, stereotaxic coordinates should be set and lowered to mark the approximate location of where a cannula should be placed. Once the burr hole is drilled, making sure that no vascular structures are damaged, dura should be cut open and gently pulled to the sides of a burr hole. A glass micropipette (1.2 mm outer diameter) with a tip diameter of 15–20 μm should be already filled with a 10 % solution of the anterograde tracer BDA (*BDA 10 kDa*) in saline, prior to lowering it to the appropriate target site (*Area A* in Fig. 1). BDA is compatible with a wide variety of delivery vehicles: distilled water, saline, or phosphate buffer (*see* Section 2).

Both iontophoretic and mechanical injections were reported for depositing BDA into the selected brain area [4]. BDA is typically iontophoretically deposited using 5 μA positive current pulses of 500-ms duration at a rate of 0.5 Hz for 30 min (see also [11, 12]). The pipette should remain in place for 60 s after the injection to minimize diffusion of the tracer along the electrode track. Upon pipette removal, skin should be cleaned and closed with sutures or surgical clips. On average, a period of 12–18 days should be allowed for *anterograde tracer transport*, from cell bodies and dendrites down the axons towards the axonal terminals (Fig. 1a). Survival time is in proportion to the length of the projection under study. Specifically, anterograde transport is estimated to span 15–20 mm of tract in 1 week [3, 4]. This anterograde tracer, BDA 10 kDa, remains stable in for up to 4 weeks in a rodent brain, and was detected up to 7 weeks following injection in a primate brain [2].

3.2 Perfusion

Following the period of survival, animals should be deeply anesthetized and transcardially perfused. Anesthesia should be effectively achieved by intraperitoneal injection of pentobarbital (70 mg/kg). If necessary, one can supplement it with an inhalational anesthetic, isoflurane. BDA tolerates a wide variety of fixatives. For optimal tissue preservation for the electron microscopic analysis, we recommend transcardiac perfusion with (1) 10 ml of heparinized saline (1000 units/ml; Elkins-Sinn, Inc., Cherry Hill, NH) at a speed of 100 ml/min, followed by (2) 100 ml of 3.75 % acrolein and 2 % paraformaldehyde in 0.1 M PB adjusted to pH 7.4 and perfused at the same speed, and (3) 100 ml of 2 % paraformaldehyde in 0.1 M PB adjusted to pH 7.4 and perfused at a speed of 100 ml/min, and an additional 100 ml of the paraformaldehyde solution perfused at a slower speed of 60 ml/min. Following perfusion, fixed brains should be removed, cut into tissue blocks of interest, and post-fixed in 2 % paraformaldehyde overnight. Finally, brain blocks should be immersed in cold 0.1 M PB and 40 μm transverse sections should be cut on a vibratome. Freely floating sections are then ready to be processed for double-labeling immunohistochemistry.

3.3 BDA Histochemistry

A method for histochemical tissue processing for the visualization of BDA was described in detail in our previous reports [6–8] and is outlined in Fig. 2. Specifically, tissue sections should be rinsed for

---

Electron Microscopic Study:
Double Labeling of BDA and TH

DAY 1: DAB visualization of BDA (brown)
- 0.1 M PB wash – 5 min
- Na-Borohydride wash – 30 min
- 0.1 M PB wash – 5 min
- TBS wash – 5 min
- **ABC Kit** (1:100; 1.5-2 hours)
- TBS wash – 2x10 min
- **DAB + H2O2** – 4 min: **BROWN REACTION**
- TBS wash – 2x10 min
- Blocking Solution for 30 min
- **Primary Ab** (mouse anti-TH; 1:1000; overnight)

DAY 2: Silver/Gold processing of TH (black)
- 0.1 M TBS wash – 3x10 min
- 0.01 M PBS wash – 3x10 min
- Blocking Solution – 10 min
- **Secondary Ab** (gold-labeled anti-mouse IgG; 1:50; 2 hours)
- Blocking Solution –10 min
- 0.01 M PBS wash – 3x10 min
- 2% Glutaraldehyde solution –10 min
- 0.01 M PBS wash – 10 min
- Citrate buffer wash – 5 min
- **Silver intensification Kit** – 15-20 min
- Citrate buffer wash – 5 min
- 0.1 M PB wash – 5 min

DAY 2: Osmification and Embedding
- 2% Osmium tetroxide in 0.1M PB – 1 h
- Dehydration in series of alcohols:
  - 30% ETOH – 3 min
  - 50% ETOH – 3 min
  - 70% ETOH – 3 min
  - 95% ETOH – 3 min
  - 100% ETOH – 2x10 min
  - Propolin Oxide – 2x10 min
- **EPON/Propolin Oxide mix (1:1)** – overnight

DAY 3: Flat Embedding
- 100% EPON – 2 hours
- Flat embedding between ACLAR Film sheets – overnight in the oven

Fig. 2 Detection of the anterograde tracer, biotinylated dextran amine (BDA), and tyrosine hydroxylase (TH) in the floating tissue sections for electron microscopic analyses. Outline illustrates steps required for double labeling of an anterograde tracer, BDA (*brown*) and noradrenergic neurons (*black*) following peroxidase (H_2O_2) reaction and silver intensification immunohistochemistry, respectively. Abbreviations for different washes and solutions are found in the *Abbreviation list* and the text of the manuscript

5 min in 0.1 M PB (pH 7.4), 30 min in 1.5 % sodium borohydride in 0.1 M PB solution, 5 min in 2–3 washes with 0.1 M PB before a final wash for 5 min in 0.1 M Tris-buffered saline (TBS, pH 7.6). Then, sections should be incubated for 1.5–2 h in a solution containing ABC Kit at room temperature, followed by two 10-min rinses in 0.1 M TBS (pH 7.6). Brown peroxidase reaction product of BDA is produced by incubation of sections for 4–5 min only, in a solution containing 22 mg of 3,3'-DAB and 20 µl of 30 % hydrogen peroxide in 100 ml of 0.1 M TBS (pH 7.6).

Considering that BDA fills the neuronal cytoplasm, the brown peroxidase staining product forms a homogeneous label inside neurons against a perfectly clear background. This characteristic makes the BDA easy to combine with other labels, either other tract tracers or additional immunohistochemistry to detect neuroactive substances (see below). In addition, the homogeneous distribution of BDA label along the entire axon allows for extremely precise light microscopic mapping of fiber tracts, organization of fascicles, as well as the study of terminal fields (Fig. 3; for an example see our previous work [13, 14].

The electron microscopy technique is a gold standard for confirming synaptic contacts between anterogradely labeled axons and neurons located in the terminal field (*see Area B* in Fig. 1a). Structural details, as seen by the electron microscope, are preserved so well that the pre- and postsynaptic membrane densities of labeled axon terminals can be clearly appreciated and distinguished from surrounding neuropil [12]. At the ultrastructural level, the immunoperoxidase detection of an anterograde tracer, BDA, can be combined with an immunoperoxidase-silver enhancement detection of a neuroactive substance present in the structures postsynaptic to tracer-labeled axon terminals (see next).

3.4 Tissue Immunohisto-chemistry

Immunogold labeling or immunogold staining is a staining technique used in electron microscopy. Gold is used for its high electron density, which increases electron scatter to give high-contrast "dark spots" or "black dots." Colloidal gold particles are most often attached to secondary antibodies, which are in turn attached to primary antibodies designed to bind a specific protein or other cell component. The electron-dense gold particle can then be seen under an electron microscope as a black dot, indirectly labeling the molecule of interest. The labeling technique can be adapted to distinguish multiple objects by using differently sized gold particles. In addition, the small gold particles have also been visualized more readily by electron microscopy after silver enhancing [15] that is utilized in the technique described here.

Tissue processing for optimization of differential peroxidase labeling and immunogold-silver immunohistochemistry with maintenance of ultrastructure in brain sections before plastic embedding was previously described in detail [16, 17]. These

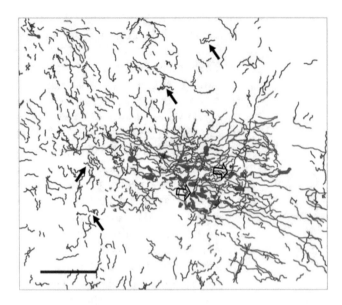

Fig. 3 This figure illustrates the relationship between BDA-labeled axons and tyrosine hydroxylase-immunoreactive, noradrenergic neurons comprising the pontine A7 cell group. This camera lucida drawing of BDA-labeled axons (*blue*) and noradrenergic profiles (neurons and dendrites shown in *red*) is representative of a transverse pontine section throughout the A7 cell group. Anterogradely labeled axons (*blue*) originate from the rostral ventrolateral periaqueductal gray where the anterograde tracer (*BDA 10 kDa*) was injected. The *open arrows* indicate examples of BDA-labeled axons closely apposed to noradrenergic somata or dendrites, while *solid arrows* indicate examples of BDA-labeled axons not in close apposition to noradrenergic profiles. Visualizing such detailed anatomical relationships between blue and red profiles was accomplished by tracing selected markers (BDA and tyrosine hydroxylase) following double-labeling immunohistochemistry. It was done directly from the light microscope under 20× objection magnification by using Neurolucida software (MBF Bioscience, Williston, VT). See also previously published work by our group [6–8, 13]. The Neurolucida software allowed for color-coding of selected profiles. Illustrated light microscopic analysis allows for excellent identification of anterograde tracer terminal fields, but is limited in its ability to visualize direct synaptic contacts. Scale bar = 200 μm

studies defined the conditions needed for optimal immunogold-silver labeling of antigens while maintaining the ultrastructural morphology of the brain sections. They also established the necessity for controlled silver intensification for both light or electron microscopic differentiation of immunogold-silver and peroxidase reaction products leading to optimal subcellular resolution. In our previous work [6–8], we combined BDA peroxidase labeling with immunohistochemistry against the enzyme responsible for synthesis of catecholamine neurons, TH. Specifically, after BDA staining (brown peroxidase reaction as described above), sections should be incubated overnight in a solution containing *PRIMARY*

ANTISERA diluted with 0.1 M TBS and contained 0.1 % BSA. After several rinses in 0.1 M TBS followed by 0.01 M phosphate buffer saline (PBS, pH 7.4), sections should be placed in a blocking solution containing 0.01 M PBS with 0.2 % gelatin (Gelatin, IGSS quality, Amersham Life Sciences, Piscataway, NJ) and 0.8 % BSA for 10 min. Sections should then be incubated for 2 h in a solution containing *SECONDARY ANTISERA*. The secondary antisera should be an antibody conjugated to 1 nm gold particles diluted with the blocking solution described above. Subsequently, sections should be rinsed again for 10 min in the blocking solution, three times for 10 min in 0.01 M PBS, and for 10 min in 2 % glutaraldehyde. After a 10-min rinse in 0.01 M PBS, sections should be washed for an additional 10 min in 0.2 M sodium citrate buffer (pH 7.4) to remove phosphate ions that could contribute to the nonspecific precipitation of silver. The optimal time for silver enhancement using the *Intense M Silver Enhancement kit* was determined to be 15–20 min. Transferring tissue sections to a 0.2 M sodium citrate buffer stops the silver enhancement reaction. This processing produces black staining of immunolabeled neurons. All incubations and washes should be carried out at room temperature with gentle agitation. For the summary of steps, *see* Fig. 2.

3.5 Tissue Embedding for Electron Microscopic Analysis

Following the silver intensification step, tissue sections should be rinsed once for 5 min in 0.1 M PB and incubated for 1 h in 2 % osmium tetroxide solution in 0.1 M PB on flat porcelain plate wells. After osmium processing, sections should be dehydrated through a series of alcohol solutions (Fig. 2) prior to undergoing tissue embedding, a final step prior to electron microscopic observation. Specifically, sections should be placed in vials containing equal parts of *propylene oxide* and *Embed 812 mixture*, and infiltrated overnight with gentle rotation at room temperature. Finally, sections should be infiltrated with *Embed 812 mixture* (without propylene oxide) for 2 h with gentle rotation before flat embedding them between two sheets of plastic. The resin should polymerize (harden) by placing the sheets in an oven at 65 °C. Once hardened, plastic-embedded tissue sections should be examined with a stereomicroscope, and small regions of interests should be trimmed out and mounted on preformed resin blocks, followed by ultramicrotome sectioning for electron microscopy.

3.6 Electron Microscopic Analyses

One should examine at least two to three 40 μm thick tissue sections per region of interest (*Area B*; Fig. 1a). Specifically, about 5–10 ultrathin sections should be cut from each 40 μm thick tissue sections with diamond knife on the ultramicrotome. Those ultrathin section (80 nm thick) should be collected on grids and counterstained with 4 % uranyl acetate and Reynold's lead citrate before being examined using a transmission electron microscope. Sections should be examined in detail and photomicrographs taken of each labeled profile (axons, dendrites, neuronal cell bodies) using a magnification of 10,000×. Finally, all the photomicrographs should be examined and all the labeled profiles classified. For the definition of cellular elements and their proper identification under electron microscope, please refer to previously published work [18, 19] or textbooks [20, 21].

3.7 Ultrastructural Characteristics of BDA Labeling

The appearance of the peroxidase reaction product in BDA-labeled profiles (axon terminals (Fig. 4), unmyelinated and myelinated axons) is distinct and described as a dense, dark flocculent label. However, its appearance can differ depending on the intensity of labeling even in the same tissue sections. Light BDA labeling allows for a clear identification of membrane profiles (e.g., asymmetric synapse; Fig. 4) and the characteristics of the synaptic vesicles (e.g., small round vs. large dense core vesicles). This is in contrast to occasional intense BDA label that completely obscures visualizations of the fine ultrastructure.

3.8 Characteristics of Silver-Intensified Gold Labeling

Immunolabeling in single thin sections should be identified by the presence of at least three gold particles within a specific cytoplasmic compartment. Therefore, lightly labeled somatic and dendritic profiles (containing less than three individual particles) should be confirmed by the presence of gold particles in at least two serial sections when possible (*see* Fig. 4). A profile that contains only one or two individual gold particles and is unlabeled in adjacent thin sections should be designated as lacking detectable immunoreactivity.

3.9 Limitation of the Technique Described

The main limitations of the present study are inherent to all anterograde transport studies and these include the *spread of the anterograde tracer* from the injection site and uptake by neurons in neighboring brain areas, as well as uptake by damaged fibers that pass through the site of the tracer deposit. To minimize this, BDA deposits should be confined to the area of interest only and not spread to the surrounding regions. Glass micropipettes with small tip diameters should be used to produce minimal neuronal damage while injecting the tracer.

An additional potentially confounding factor is the *retrograde labeling* of neurons with axons that project to the tracer injection site, and subsequent anterograde transport in axon collaterals [3, 12, 23]. Although a systematic study of this retrograde–anterograde transport was not done, only a small number of retrogradely labeled neurons are observed when using BDA 10 kDa. These are usually randomly distributed are not concentrated in any particular region. It is unlikely that these scattered retrogradely labeled neurons provide a significant contribution to the labeling of axons and terminals seen in the area of interest. In summary, one should pay

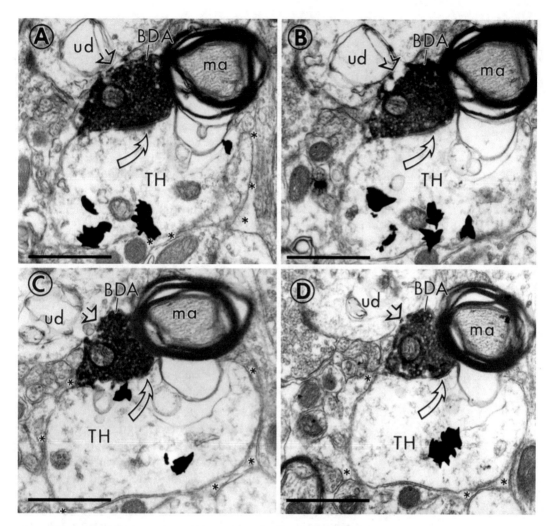

Fig. 4 Ultrastructural illustration of a single anterogradely labeled terminal forming a synapse with a noradrenergic dendrite (**a–d**). Specifically, these four serial electron micrographs illustrate an example of a synaptic contact (*large curved arrow*) between a densely anterogradely labeled axon terminal (BDA), originating from the neurons in the ventrolateral periaqueductal gray, and a noradrenergic dendrite labeled by tyrosine hydroxylase (TH). Note the difference in labeling for electron microscopy between peroxidase label of BDA and silver intensified immunogold label (particulate matter) of TH. This figure illustrates two additional points. The first one is related to the importance of identifying double labeling in serial thin sections. A profile containing only one (**d**) or two (**c**) immunogold particles could not be designated as noradrenergic unless the adjacent thin sections identify more than two gold particles in the same profile (**a** and **b**). Furthermore, characteristics of an asymmetric synapse are most clearly seen in *Panel* (**a**). Asymmetric synapses are identified by the presence of thick postsynaptic densities (Gray's Type I), while symmetric synapses have thin densities (Gray's Type II) located both pre- and postsynaptically [22]. Although *panels* (**b**) and (**d**) clearly show the membrane structures between the two profiles, unequivocal characteristic of an asymmetric synapse (postsynaptic density) is missing. Similarly, the densely labeled axon terminal (BDA) is closely apposed (*small open arrow*) to an unlabeled dendrite (ud), but synaptic specializations are not readily apparent in any of the four serial sections. Scale bars = 1 μm. *ma* myelinated non-labeled axon. *Panels* (**a**) and (**b**) are used with permission from *Journal of Comparative Neurology*

attention to the extent of retrograde labeling, if any (e.g., BDA-labeled neurons and dendrites in the area of ultrastructural analysis). In addition, an anterograde tracer *BDA 10 kDa* (Fig. 1a) should not be interchanged with another member of the dextran amine tracers, *BDA 3 kDa*. The latter one is better suited for retrograde tracing purposes [4]. This smaller molecule is transported in a retrograde fashion, from the neuronal terminals back to the neuronal cell body. For schematic illustration of principle of retrograde transport, refer to Fig. 1b. Future studies should consider a novel anterograde tracing technique involving a GFP-containing recombinant adeno-associated virus vector that virtually has no retrograde tracing but has the same pattern of distribution as biotinylated dextrans and another anterograde tracer, Phaseolus vulgaris leucoagglutinin [24, 25].

4 Notes

4.1 Anterograde Tracer Injection

1. During stereotaxic injections, the size/weight of the animal should be taken into account. As the animal grows, the coordinates might change, especially if one is trying to inject BDA into a small brain nucleus. Thus, several attempts should be made initially until ideal stereotaxic coordinates are set for an animal of a particular size.

2. Micropipette tip diameter may play a significant role in the amount of BDA deposited and thus the amount of tracing that will be detected. If the tip is inadvertently broken, the injection might be quite significant. One should examine the tip of the micropipette upon the completion of injection to confirm the tip was intact. Detection of a broken tip can subsequently explain the larger than expected BDA injection site.

3. If large volumes of BDA tracer are pressure injected, one should be aware of the potential uptake of the BDA tracer by the fibers of passage that can lead to erroneous tracing, including a retrograde labeling (*see* below; Fig. 1b).

4.2 Perfusion and Vibratome Sectioning

1. For the most successful transcardiac perfusion, one would ideally want to place the perfusion cannula into the left heart ventricle, and cut a hole in the right atria for the perfusate to escape. Occasionally, the cannula is placed into the right ventricle, instead of the left, leading to primary infusion of lungs. Lungs then become inflated and filled with solution. One should try to reposition the cannula into the left ventricle. Irrespectively, perfusion should continue, as the brain will eventually be fixed even if the perfusate is injected into the pulmonary circulation first.

2. Tissue blocks should be embedded into agar (6 %) dissolved in water to create agar gel. Once firm, sides of the agar should be trimmed and tissue block should be glued to the glass surface of the vibratome platform using an Instant Krazy glue (Cambridge, MA). It is nice to prepare ice-cubes out of the 0.1 M PB to be used in case the vibratome well solution starts to warm up. Vibratome sectioning at room temperature is imperative for good tissue preservation prior to electron microscopic analysis. Freezing the tissue will break the membranes and destroy the fine-tissue morphology.

3. Each time a new tissue block is cut, one should use a new blade.

4. Finally, before starting the sectioning, one should notch the CONTRALATERAL SIDE to the injection site of the tissue block using an injecting needle. Nothing is more imperative for electron microscopic study of tracers than knowing the side of a tracer injection.

4.3 BDA Histochemistry

1. One could use light microscopic visualization of only BDA to confirm adequate location of BDA injection sites. This is especially convenient when the injection site (Area A, Fig. 1a) is far from the area of analysis (Area B, Fig. 1a). Tissue sections containing the injection site should be collected and processed separately following the protocol described above. For the light microscopic analysis, one could also add Tris/Triton-X 100 solution wash for 20 min, followed by Tris/saline wash prior to starting the ABC reaction. Note that Triton-X 100 is not necessary for BDA labeling and that any processing of tissue with Triton-X 100 would be detrimental for electron microscopic studies since all the protein structures would be destroyed.

2. One should consider light microscopic analysis of an anterograde tracer double-labeled with a neurochemical marker as a prerequisite for ultrastructural analyses. This is to actually show that BDA-anterogradely labeled axons are indeed present in the area of interest. Light microscopic analysis is quite limited in its resolution; light microscopic results only imply close appositions between anterogradely labeled axons (originating from the Area A, Fig. 1a) and neurons of interest (Area B, Fig. 1b; e.g. noradrenergic neurons of the pontine A7 cell group; Fig. 3).

4.4 Tissue Immunohisto-chemistry

1. When considering multiple immunostaining protocols, biotinylated secondary antibodies might cross-react with the BDA during the detection step for the additional markers. Therefore, biotinylated secondary antibodies should be avoided when double labeling also involves BDA. Note that for the described protocol to work for electron microscopic analysis, secondary antibodies must be conjugated to gold particles.

2. For an overview and examples of peroxidase substrates currently available that could be combined for double tract tracing paradigms, refer to a recent review [2]. Report by Anderson et al. [26] described successful triple labeling using different substrates (DAB as flocculent precipitate, silver-intensified immunogold, and benzidine hydrochloride (crystalline texture)).

3. For additional examples of double-labeling immunohistochemistry for electron microscopic techniques, see studies that combine pre-embedding peroxidase detection of anterograde tracer with post-embedding GABA immunohistochemistry to determine the GABA presence inside the tracer-containing presynaptic axon terminals [27–29].

4.5 Tissue Embedding for Electron Microscopic Analysis

1. Antibodies and gold particles cannot penetrate the resin used to embed samples for imaging. Labeling prior to embedding the sample (as described in previous sections) can reduce the negative impact of this limitation.

4.6 Approach for Electron Microscopy

1. Immunogold labeling can introduce artifacts, as the gold particles reside some distance from the labeled object. Very thin sectioning is required during sample preparation; thus 80 nm thick sections should be cut with ultramicrotome.

2. Electron microscopic photomicrographs should be exclusively prepared from regions of tissue located only near the surface of the sections. Remember that the tissue sections are 40 μm thick and that immunolabeling penetrates and labels the profiles of interest only on the surfaces of each section (see Fig. 4). Limitation of this immunohistochemical method to detect trace amounts of labeling may contribute to an underestimation of the number of profiles detected by silver-intensified gold labeling. This potential limitation can therefore be minimized by not only collecting tissue sections near the tissue surface but also by ensuring that both labels were clearly present in fields of analysis.

5   Conclusions

The significance of this technique, double labeling of anterograde tracer (BDA 10 kDa) combined with silver-intensified gold labeling immunohistochemistry, is several. Visualization of synapses between the neurons of a particular neuronal circuit provides direct ultrastructural evidence of neuronal circuit communication (axon-to-dendrite communication; monosynaptic projection). Furthermore, visualization of synapses of anterograde tracer with other axons in the area of interest (axo-axonic synapses) provides supportive evidence of indirect influence anterogradely labeled axons exert on a synaptic transmission. Finally, establishing anatomical evidence of neuronal circuits is a prerequisite to subsequent physiological studies.

Acknowledgements

This research was supported by (1) USPHS Grant DAO3980 from the National Institutes on Drug Abuse (NIDA) to Dr. Herbert K. Proudfit; (2) grants ANA#96002249 and NIDA 09082 to Dr. Elisabeth J. Van Bockstaele; as well as the (3) National Institutes of Health R03 DA030874 grant to Dr. Dusica Bajic.

References

1. Elias LJ, Saucier DM (2005) Neuropsychology: clinical and experimental foundations. Pearson, London, p 560

2. Lanciego JL, Wouterlood FG (2011) A half century of experimental neuroanatomical tracing. J Chem Neuroanat 42(3):157–183

3. Veenman CL, Reiner A, Honig MG (1992) Biotinylated dextran amine as an anterograde tracer for single- and double-labeling studies. J Neurosci Methods 41(3):239–254

4. Reiner A et al (2000) Pathway tracing using biotinylated dextran amines. J Neurosci Methods 103(1):23–37

5. Reiner A, Honing MG (2006) Dextran amines: versatile tools for anterograde and retrograde studies of nervous system connectivity. In: Zaborszky L, Wouterlood FG, Lanciego JL (eds) Neuroanatomical tract-tracing 3: molecules, neurons, and systems. Springer, New York City, NY, pp 304–335

6. Bajic D, Proudfit HK, Van Bockstaele EJ (2000) Periaqueductal gray neurons monosynaptically innervate extranuclear noradrenergic dendrites in the rat pericoerulear region. J Comp Neurol 427(4):649–662

7. Bajic D, Van Bockstaele EJ, Proudfit HK (2001) Ultrastructural analysis of ventrolateral periaqueductal gray projections to the A7 catecholamine cell group. Neuroscience 104 (1):181–197

8. Bajic D, Van Bockstaele EJ, Proudfit HK (2012) Ultrastructural analysis of rat ventrolateral periaqueductal gray projections to the A5 cell group. Neuroscience 224:145–159

9. Bozzola JJ, Russell LD (1999) Specimen preparation for transmission electron microscopy. In: Bozzola JJ, Russell LD (eds) Electron microscopy : principles and techniques for biologists. Jones & Bartlett Publishers, Sudbury, MA, pp 16–47

10. Hayat MA (2000) Principles and techniques of electron microscopy: biological applications. Cambridge University Press, New York, NY

11. Gonzalo N et al (2001) A sequential protocol combining dual neuroanatomical tract-tracing with the visualization of local circuit neurons within the striatum. J Neurosci Methods 111 (1):59–66

12. Wouterlood FG, Jorritsma-Byham B (1993) The anterograde neuroanatomical tracer biotinylated dextran-amine: comparison with the tracer Phaseolus vulgaris-leucoagglutinin in preparations for electron microscopy. J Neurosci Methods 48(1-2):75–87

13. Bajic D, Proudfit HK (2013) Projections from the rat cuneiform nucleus to the A7, A6 (locus coeruleus), and A5 pontine noradrenergic cell groups. J Chem Neuroanat 50–51:11–20

14. Bajic D, Proudfit HK (1999) Projections of neurons in the periaqueductal gray to pontine and medullary catecholamine cell groups involved in the modulation of nociception. J Comp Neurol 405(3):359–379

15. Scopsi L et al (1986) Silver-enhanced colloidal gold probes as markers for scanning electron microscopy. Histochemistry 86(1):35–41

16. Chan J, Aoki C, Pickel VM (1990) Optimization of differential immunogold-silver and peroxidase labeling with maintenance of ultrastructure in brain sections before plastic embedding. J Neurosci Methods 33 (2-3):113–127

17. Pickel VM, Chan J (1993) Electron microscopic immunocytochemical labeling of endogenous and/or transported antigens in rat brain using silver-intensified one-nanometer

colloidal gold. In: Cuello AC (ed) Immunohistochemistry II. Willey, New York City, NY, pp 265–280

18. Peters A, Palay SL (1996) The morphology of synapses. J Neurocytol 25(12):687–700

19. Peters A, Palay SL, Webster HD (1992) Fine structure of the nervous system: neurons and their supporting cells, 3rd edn. Oxford University Press, New York, NY

20. Hunter EE, Silver M (1993) Practical electron microscopy: a beginner's illustrated guide, 2nd edn. Cambridge University Press, New York, NY, p 188

21. Bozzola JJ, Russell LD (1998) Electron microscopy, 2nd edn. Jones & Bartlett Publishers, Toronto, ON

22. Gray EG (1959) Axo-somatic and axo-dendritic synapses of the cerebral cortex: an electron microscope study. J Anat 93:420–433

23. Brandt HM, Apkarian AV (1992) Biotin-dextran: a sensitive anterograde tracer for neuroanatomic studies in rat and monkey. J Neurosci Methods 45(1-2):35–40

24. Chamberlin NL et al (1998) Recombinant adeno-associated virus vector: use for transgene expression and anterograde tract tracing in the CNS. Brain Res 793(1-2):169–175

25. Gautron L et al (2010) Identifying the efferent projections of leptin-responsive neurons in the dorsomedial hypothalamus using a novel conditional tracing approach. J Comp Neurol 518 (11):2090–2108

26. Anderson KD, Karle EJ, Reiner A (1994) A pre-embedding triple-label electron microscopic immunohistochemical method as applied to the study of multiple inputs to defined tegmental neurons. J Histochem Cytochem 42(1):49–56

27. Freund TF, Antal M (1988) GABA-containing neurons in the septum control inhibitory interneurons in the hippocampus. Nature 336 (6195):170–173

28. Omelchenko N, Bell R, Sesack SR (2009) Lateral habenula projections to dopamine and GABA neurons in the rat ventral tegmental area. Eur J Neurosci 30(7):1239–1250

29. Omelchenko N, Sesack SR (2009) Ultrastructural analysis of local collaterals of rat ventral tegmental area neurons: GABA phenotype and synapses onto dopamine and GABA cells. Synapse 63(10):895–906

Neuromethods (2016) 115: 81–103
DOI 10.1007/7657_2015_97
© Springer Science+Business Media New York 2015
Published online: 05 February 2016

Three-Dimensional Electron Microscopy Imaging of Spines in Non-human Primates

R.M. Villalba, J.F. Paré, and Y. Smith

Abstract

Dendritic spines are the main sites of excitatory glutamatergic synapses in the central nervous system. Morphological, ultrastructura l, and numerical changes in dendritic spines are associated with long-term potentiation or depression of normal synaptic transmission, and with various brain diseases and pathological conditions that affect glutamatergic transmission. Thus, a deep understanding of the structural changes that affect dendritic spines in normal and pathological conditions is a key element of structure-function relationships that regulate synaptic transmission in the mammalian brain. In this chapter, we describe the procedure used in our laboratory that combines immuno-electron microscopy methods (to identify specific populations of presynaptic terminals or dendritic spines), serial ultrathin sectioning, and three-dimensional electron microscopy reconstruction to analyze ultrastructural and morphometric changes of individual dendritic spines in rhesus monkey models of brain diseases, most specifically related to Parkinson's disease.

Keywords: Non-human primates, Brain issue, Perfusion, Tissue processing, Immunocytochemistry, Ultramicrotomy, Serial sections, Transmission electron microscopy, Dendritic spines, 3D reconstruction, Quantitative analysis

1 Introduction

In 1891, Ramon y Cajal provided the first detailed description of dendritic spines as small protrusions on the surface of neurons [1]. Since then, studies of spine size, morphology, number, morphogenesis, and plasticity have been at the forefront of research in various laboratories.

Although dendritic shafts and spines could be seen at the light microscopic level, it wasn't until the 1950s that transmission electron microscopy (TEM) allowed to demonstrate that dendritic spines are the primary sites of excitatory synapses in the CNS [2, 3]. The use of three-dimensional (3D) reconstruction of individual spines through serial ultrathin sections has since been produced, initially by tracing the outline of the structures of interest from illuminated EM images on a sheet of acetate [4]. Over the years, 3D EM reconstruction methods and analyses have relied increasingly on computers, digital cameras [5–7], and computerized software tools such as *Reconstruct* (http://synapes.bu.edu) [8–10] or *Midas* (*manual image alignment for MRC files*, http://bio3d.colorado.edu/imod/) [11] to register images taken from serial ultrathin sections with TEM.

The use of computer-assisted 3D reconstructions of individual spines at the electron microscopic level has been applied to the analysis of individual dendritic spines from specific neuronal populations (i.e., hippocampal and cortical pyramidal neurons, cerebellar Purkinje cells, striatal medium spiny neurons, etc....) under physiological and pathological conditions. Together, these studies provided evidence that spines are highly plastic entities that are capable of complex structural remodeling in response to physiological or pathophysiological alterations, and that these structural changes are correlated with altered synaptic transmission in normal and diseased states [12–31].

In recent years, our laboratory has devoted significant interest in characterizing the ultrastructural changes of dendritic spines and related glutamatergic synapses on striatal neurons in the non-human primate model of Parkinson's disease. We have combined immuno-electron microscopy procedures (to identify specific populations of presynaptic terminals), serial ultrathin sectioning, and 3D EM reconstruction to quantitatively assess and compare various ultrastructural parameters known to be closely linked with synaptic strength and efficacy of glutamatergic synapses in the striatum of adult rhesus monkeys between normal and parkinsonian states. More specifically, these approaches allowed us to quantify and compare various structural parameters of striatal spine morphology and specific glutamatergic synapses between normal monkeys and animal models of Parkinson's disease. Thus, using striatal projection neurons as a working model, the following protocol will describe the technical procedures used in our laboratory to generate series of high-quality EM images that can be used for quantitative analysis to determine ultrastructural changes in the morphology, synaptic connectivity, and perisynaptic glial coverage of axo-spinous cortical or thalamic synapses between normal and 1-methyl-4-phenyl-1,2,3,6-tetrahydropyridine (MPTP)-treated parkinsonian monkeys [26–29].

2   Materials

2.1   General Reagents, Equipment, and Supplies

pH meter, hot plate stirrer, forceps, transfer pipettes, Eppendorf pipettes®, pipette tips, beakers (glass and plastic), graduated cylinders (glass and plastic), flasks, conical flasks, petri dishes (glass and plastic), scintillation vials, Erlenmeyer flasks, culture plates (24- and 6-well plates), distilled water apparatus, gloves, small paint brushes, filter paper, microscope slides, cover glasses, safety goggles, lab coats, glass markers, spatulas, fume hood, dissector scope, microscope, watch glasses, magnetic stirring bars, digital scale balance, weight boats and dishes, funnels (glass and plastic), scissors, safety razor blades, plastic syringes, syringe filters, Kimwipes®, Parafilm®, aluminum foil, sharp safety containers, cyanoacrylate glue.

2.2 Buffers and Solutions

1. Phosphate buffer (PB) 0.2 M pH 7.4:

   Solution A: 27.6 g sodium phosphate monobasic in 1 liter (1 l) distilled water. $NaH_2PO_4 \cdot H_2O$. Store it in the fridge (4 °C).

   Solution B: 53.6 g sodium phosphate dibasic heptahydrate in 1 l distilled water. $Na_2HPO_4 \cdot 7H_2O$

   Mix 800 ml Solution B with 200 ml Solution A for a total volume of 1 l or 5.52 g $NaH_2PO_4 \cdot H_2O$ + 42.88 g $Na_2HPO_4 \cdot 7H_2O$ in 1 l. Adjust the pH to 7.4 with NaOH or HCl. Store it in the fridge (4 °C).

2. PB (0.1 M pH 7.4): Mix 400 ml Solution B with 100 ml Solution A in 500 ml distilled water, or 2.76 g $NaH_2PO_4 \cdot H_2O$ + 21.44 g $Na_2HPO_4 \cdot 7H_2O$ in 1 l. Adjust the pH to 7.4 with NaOH or HCl. Store it in the fridge (4 °C).

3. PBS (0.01 M pH 7.4): Use working solution from PBS in fridge or use stock solution 1:10 in distilled water. Adjust the pH to 7.4 with NaOH or HCl and store in the fridge (4 °C). For a 10× recipe: Dissolve in 20 l of distilled water: 428.8 g of sodium phosphate dibasic ($Na_2HPO_4 \cdot 7H_2O$), 55.2 g of sodium phosphate monobasic ($NaH_2PO_4 \cdot H_2O$), 1800 g of NaCl. Adjust pH and leave at room temperature (RT).

4. TRIS (0.05 M pH 7.6): Dissolve 6.05 g THAM ($C_4H_{11}NO_3$) in 1 l distilled water. Adjust the pH to 7.6 with NaOH or HCl. Store in the fridge (4 °C).

5. Antifreeze solution: To prepare 1 l: 13.75 g sodium phosphate monobasic (solution A) $NaH_2PO_4 \cdot H_2O$, 25.75 g sodium phosphate dibasic heptahydrate (solution B) $Na_2HPO_4 \cdot 7H_2O$; 400 ml distilled water, 300 ml ethylene glycol, and 300 ml glycerol.

6. Cryoprotectant solution: To prepare 1 l: Mix 200 ml PB (0.2 M pH 7.4) with 520 ml distilled water, 80 ml glycerol, and 200 g sucrose. This solution is used before freezing the sections that are going to be processed for electron microscopy.

7. Pioloform solution: Under the hood, dilute 0.75 g of Pioloform in 100 ml of chloroform and stir vigorously (the final volume of the solution must be 35 ml). Mix 15 ml of Pioloform with 20 ml of chloroform and filter it in a glass staining dish and cover it. Store the solution in the fridge after use. This solution

needs to be removed from the fridge 3 h before the beginning of the technique.

2.3 Perfusion and Tissue Fixation

1. Large fume hood or downdraft table: Perfusion procedures must be performed under either a fume hood or on a downdraft table to reduce exposure to aldehyde vapors.

2. Personal protective equipment: Eye protection, face shields, masks, gloves (double), surgical gown with long sleeves and elastic cuffs, shoe covers, head cover. Any contact with non-human primate tissue or liquids must be done in accordance with the Standing Operating Procedures in place at your institution.

3. Perfusion/dissection aid tools: Dissecting scissors, bone cutters, retractors, forceps, hemostats, scalpel and blades, spatulas, bone rongeurs, bone saw, peristaltic perfusion pump.

4. Animals: Adult rhesus monkeys (*Macaca mulatta*).

5. Anesthetics: Pentobarbital (100 mg/kg i.v.).

6. Oxygenation system: O_2/CO_2 (95/5 %) gas tank with regulator, assorted tubing, connectors, and clamps.

7. Ringer's solution: In 1 l distilled water, dissolve the following reagents: 0.60 g HEPES, 11.86 g NaCl, 0.223 g KCl, 0.353 g CaCl, 2.18 g $NaHCO_3$, 0.177 g KH_2PO_4, 0.32 g $MgSO_4$, 1.8 g D-glucose (*see* **Note 1**).

8. Fixative: Whatever method of fixation is selected, it must serve the dual purpose of retaining the essential structural and antigenic components of the tissue, without introducing any material that may interfere with the overall quality of the immunohistochemical labeling. In our case, we use 2 l of a mix of paraformaldehyde (4 %) and glutaraldehyde (0.1 %) to perfuse an adult rhesus monkey. To prepare 2 l of 4 % paraformaldehyde and 0.1 % glutaraldehyde fixative: Heat 1 l distilled water until the temperature reaches 60 °C, then dissolve 80 g of paraformaldehyde with strong stirring action for 20 min and add NaOH drop by drop until the solution is almost clear. Let the solution cool for 1 h, filter it in another beaker, and add 1 l PB (0.2 M pH 7.4). For a 0.1 % concentration in 2 l of fixative, substitute 8 ml of fixative with 8 ml of 25 % glutaraldehyde solution (*see* **Notes 2–6**).

2.4 Tissue Sectioning

1. Vibratome: See: http://physics.ucsd.edu/neurophysics/Manuals/tpi/Vibratome1000.pdf

2. PBS or antifreeze solution.

3. 24-well plates.

4. Petri dishes.

5. Dissecting microscope.

6. Safety razor blades.

2.5 Immunostaining

1. Primary and secondary antibodies.

2. VECTASTAIN® Elite® ABC System.

3. 3,3′-diaminobenzidine tetrahydrochloride (DAB; Sigma, St. Louis, MO) solution: Use a disposable polypropylene beaker to prepare this solution: 50 ml of TRIS buffer (0.05 M pH 7.4); 0.5 ml of imidazole (from a 1.0 M solution), and 0.0125 g of DAB (*see* **Note 7**).

2.6 Electron Microscopy (EM) Tissue Processing

1. Osmium tetroxide (1 %): Dilute 5 ml osmium (sold in 4 % solution) with 15 ml PB (0.1 M pH 7.4) (*see* **Note 8**).

2. Ethanol.

3. Uranyl acetate.

4. Durcupan ACM mixture (Fluka, Durcupan ACM; Fluka, Fort Washington, PA): Prepare the resin as follows: 10 g Component A, 10 g Component B, 0.3 g Component C, 0.2 g Component D. Mix the resin with a wood stick in a polypropylene beaker (*see* **Note 9**).

5. Aluminum dishes, glass slides and mineral oil.

6. Laboratory oven.

2.7 EM Grids

1. Grids: Single slot (oval hole).

2. Pioloform-coated copper grids:

   Coating grids (*see* **Note 10**): Fill the glass container with water (you should put the glass container in a plastic tray). Before spreading any membrane on the water, remove dust or residues by passing a pipette on the surface. Dip a clean, uncoated slide into the pioloform solution and let it dry a few seconds. Pass a single edge razor blade along the slide with an angle of 45° and on the higher and lower part of the slide. Repeat procedure for both sides of the slide. Dip the slide very slowly in the water and the membranes should come off. Verify the thickness here and check for dust residues (*see* **Note 11**). Put the grids on the membrane so that the polished side of the grids face the membrane, and let the end of the membrane free of grids to facilitate the recuperation.

3. Put a piece of parafilm on a clean slide and dip it in the water so that the slide touches the empty end of the membrane first. Remove the parafilm piece from the slide and put it down in a Petri dish (*see* **Note 12**).

2.8 Serial Ultrathin Sections and Staining (Fig. 1)

1. Ultramicrotome: Ultracut T2; Leica Germany (Fig. 1a, d, e).

2. Diamond Knifes: (Fig. 1b)

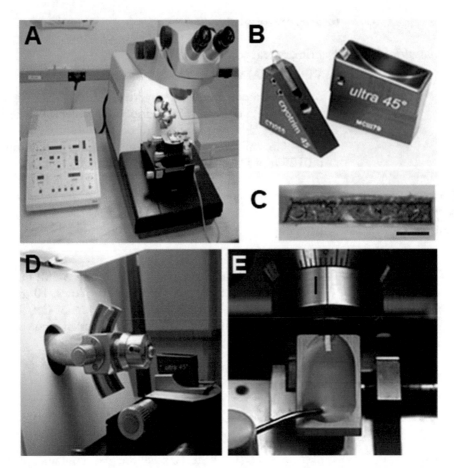

Fig. 1 Ultramicrotome, diamond knives and tissue blocks. (**a**) Image on an ultramicrotome used to prepare semithin (1–2 µm) and ultrathin (60–70 nm) sections. (**b**) Diamond knives used for trimming tissue blocks with top and bottom parallel faces (cryotrim 45°), and to produce ultrathin sections (ultra 45°) with smooth sample surfaces, good ultrastructure, and very regular thickness. Notice that in the ultrathin 45° diamond knife, the diamond is mounted in a precision made metal holder called boat. (**c**) Light microscope image of a tissue block trimmed and ready to be cut in ultrathin sections. (**d**) Lateral view of the diamond knife, the diamond knife holder and the ultramicrotome arm holding the trimmed tissue block. (**e**) Top view of the diamond knife and ultrathin serial sections forming a ribbon on the water surface in a diamond knife boat. Scale Bar in (**c**) =125 µm

> Trimming: DiATOME Cryotrim 45°
>
> Sectioning: DiATOME Ultrathin 45°

3. Serial sections manipulation and storage:

   EM Grid boxes:

   – Grid storage boxes.

   – Grid staining matrix system (cover and body) and vessels (PELCO®): Reduces the chance of mechanical damage, ensures equal staining, and rinses time.

Fine forceps, eyelash tool, plastic syringes, and filters.

4. Lead Citrate (*see* **Note 13**): Weigh 0.266 g of lead nitrate. Dissolve it in a vial containing 9.6 ml of distilled water and mix thoroughly until it is completely dissolved. Weigh 0.354 g of sodium citrate and dissolve carefully. At this point, the solution should have a milky look. Add 0.4 ml of NaOH (5 N) and mix carefully. The milky look should disappear immediately after NaOH addition. If kept in the fridge (4 °C), this solution is good for 1 week.

2.9 Serial Section Imaging and Analysis

1. Zeiss EM-10C and Jeol electron microscopes with a CCD camera (DualView 300 W).

2. Workstation running RECONSTRUCT™ for imaging analysis and 3D reconstruction.

   Available at: http://synapses.clm.utexas.edu/

3 Methods

3.1 Perfusion and Tissue Fixation

All monkey perfusions are performed in the necropsy room in the presence of an attending veterinarian and a pathologist or lab staff member. The Emory SOP guidelines for monkey euthanasia are followed during these procedures.

1. Monkeys prepared for perfusion are first sedated with ketamine (10 mg/kg i.m.) in their home cage, transferred to the procedure room, and undergo a tracheal intubation to allow bag ventilation during perfusion. The animals are then brought to the necropsy room, given an i.v. injection of heparin, and deeply anesthetized with an overdose of pentobarbital (100 mg/kg, i.v.). The level of anesthesia is assessed with a toe pinch.

2. Oxygenate the Ringer's solution (oxygenated, O_2/CO_2; 95 %/5 % if possible) for about 5 min prior to start of perfusion. Then, flush it through the tubing connected to the peristaltic pump. To avoid formation of air bubbles, let the pump work at a low rate (~20 ml/min). Use a 14-gauge × 1 ½″ needle to perfuse adult rhesus monkeys. The size of the needles can be reduced if infants or smaller monkey species are used.

3. When the animal is deeply sedated, open the thoracic cage to access the heart and start the perfusion. Before piercing the heart, clamp the descending aorta, open the right atrium and penetrate the left ventricle with perfusion needle, use a hemostat to clamp the needle in the left ventricle.

4. Increase the speed of the pump to ~100 ml/min and rapidly flush the vasculature with about 200–400 ml of Ringer's

solution for an adult (~5–10 kg) rhesus monkey. The volume can be reduced for smaller animals. Make sure there is no air bubble in the solution along the plastic tube.

5. Stop the pump, and switch the tube to the fixative solution. Restart the pump to perfuse about 500–700 ml of the fixative at ~100 ml/min flow rate. Then, lower the speed to ~50 ml/min and perfuse the rest of the solution over an additional 30–40 min.

6. At the end of the perfusion, the head of the animal is removed and the top of the skull opened using the bone cutters or the saw blade. Once the brain is exposed, remove the dura mater, fix the head in a stereotaxic frame, and cut the brain into 10-mm-thick blocks in the frontal plane. Gently remove blocks of brain from the skull and post-fix them in 4 % paraformaldehyde (no glutaraldehyde) overnight. Then transfer the blocks to PBS (0.01 M, pH 7.4), until they are sectioned with a vibrating blade microtome (60 μm thick).

3.2 Pre-embedding Immunocytochemistry Protocol

1. Vibrating blade microtome: Prior to sectioning, it is important to trim the top surface of the specimen with the sectioning blade. When the specimen is flat and the mounting block is clamped into the specimen vise in the desired orientation relative to the blade advance, the top surface of the specimen should be kept approximately horizontal.

2. The specimen should be raised (or lowered) to a position just below the blade edge. The speed and the amplitude settings should be initially set to the "0" position. The advance speed should be at low setting, while the amplitude should be at medium to high setting. The section thickness should be incremented at 60–70 μm intervals.

3. The 60-μm-thick serial sections are transferred using a small brush from the specimen bath to 24-well plates filled with PBS (0.01 M, pH 7.4) or antifreeze solution (if sections are going to be stored in −20 °C freezer) (*see* **Note 14**).

4. After completion of sectioning, the specimen bath, specimen vise, and sectioning blade holder should be cleansed with distilled water. For a detailed info: http://physics.ucsd.edu/neurophysics/Manuals/tpi/Vibratome1000.pdf

5. The Vibratome coronal sections (60 μm) containing the region of interest (ROI) is then selected and processed for pre-embedding immunohistochemical staining. The total number of sections to be processed varies according to your needs and goals of the study.

6. Sections are placed in a net with PBS, and then in a sodium borohydride solution (1 % in PBS) under the hood (for 20 min), and rinsed in PBS until no bubbles remain on the tissue (usually 4–5 times) (*see* **Note 15**).

7. Sections are transferred to the cryoprotectant solution (100 %) at RT (for 20 min). The net is then taken out of the solution and put onto a paper towel to remove the excess cryoprotectant before being placed in a −80 °C freezer (20 min). Finally, the net with the tissue is transferred back in the 100 % cryoprotectant for 10 min at RT followed by immersion in 70 and 50 % cryoprotectant solution, and PBS (10 min in each at RT).

The following incubations are done in 6-well plates and with agitation.

8. Sections must first be pre-incubated in a PBS solution containing 1 % normal serum (usually from the species that generates the secondary antibody) and 1 % bovine serum albumin (BSA) for 60 min at RT.

9. This is followed by incubations in the primary antibody solution prepared as follows: PBS containing 1 % normal serum, 1% bovine serum albumin (BSA), and the needed quantities of primary antibodies solution for optimal dilution. The incubation time is usually 48 h at 4 °C. Once completed, this incubation is followed by rinses in PBS (usually 3 × 10 min).

10. Sections are then incubated in the biotinylated secondary antibodies solution (often diluted at ~1:200) with PBS containing 1 % normal serum and 1 % BSA. The incubation time of this reaction is 90 min at RT. Once completed, sections are rinsed in PBS for 30 min (3 × 10 min). Place sections in the avidin-biotin-peroxidase complex (ABC) solution for 90 min at RT. This solution needs to be prepared 30 min in advance according to the vendors' instructions. The optimal dilution (~0.2–1 % in PBS/BSA) must be adjusted for the different antibodies under study. Sections are then washed in PBS (2 × 10 min), followed by a TRIS buffer (0.05 M pH 7.6) rinse for 10 min before the DAB reaction.

11. DAB reaction: Just before the reaction, add to 50 ml of DAB solution (0.0125 g of DAB in 50 ml of TRIS), 1 ml of hydrogen peroxide (H_2O_2 0.3 %).

12. Sections are then transferred into a 6-well plate containing a solution of 0.01 M imidazole, 0.005 % hydrogen peroxide, and 0.025 % DAB. After 10 min, this reaction is stopped by rinses (5 × 1 min) in PBS (0.01 M, pH 7.4).

3.3 EM Tissue Processing

1. Osmication (*see* **Notes 16** and **17**): Sections are washed (3 × 5 min) in PB (0.1 M pH: 7.4) and then transferred with a small brush to watch glasses (very important that sections do

not fold over before contact with the osmium). Remove the excess of PB (0.1 M pH: 7.4) with a Pasteur pipet. Add osmium (1 %) drop by drop and wait 20 min. Take out the osmium and rinse (1 × 2 min + 1 × 5 min) with PB (0.1 M pH 7.4). With a small brush, remove the osmicated sections out of the watch glasses, place them back in the 6-well plate, rinse (2 × 5 min) with PB (0.1 M pH 7.4).

2. Dehydration and uranyl acetate staining (*see* **Note 18**): Uranyl acetate (1 %) needs to be prepared in 70 % alcohol, and stored in a dark/opaque glass container (cover the container with aluminum foil, if necessary). This solution takes 15 min to dissolve and must be filtered before use. Once the uranyl acetate solution is ready, the sections are dehydrated in ascending concentrations of ethanol solutions: 15 min in 50 % alcohol, 35 min in 1 % uranyl acetate solution (in 70 % alcohol in the dark), 15 min in 90 % alcohol, 2 × 10 min in 100 % alcohol, and 2 × 10 min in propylene oxide (*see* **Note 19**).

3. Embedding in Epoxy Resin (Durcupan ACM) (*see* **Note 20**): After dehydration with ethyl alcohol and propylene oxide, the tissue is ready for resin embedding. Once the resin components have been mixed according to the vendor's instructions (see Materials), pour the solution in aluminum dishes, and use a fine brush to put down the sections one at a time in the resin until they are completely covered. Sections must be handled gently during this process because they are very fragile. Leave sections in the resin for at least 12 h at RT under the hood.

4. Mounting/Flat-embedding: Sections are then taken out of the resin, mounted on glass slides, and coverslipped. The slides and cover glasses used to mount the sections need to be rubbed with a thin layer of mineral oil before they come in contact with the resin. The aluminum dishes containing the sections are placed one at a time on a hotplate set at 50 °C. Once the resin has slightly liquefied, the sections are transferred from the dish to the oily slides using a thin brush. Cover the sections with an oily cover glass and spread the resin by pressing carefully on the cover glass; this will force the air bubbles to go out and assure that sections are completely covered with a thin layer of resin. The resin-embedded sections are then placed in the oven at 60 °C for 48 h (*see* **Notes 21** and **22**).

5. Re-embedding (*see* **Note 23**): After the resin has polymerized, the sections are taken out from the oven and the coverslips removed using a single edge razor blade by slipping it between the resin and cover glass. The ROI is localized using the stereomicroscope, and then the slide is placed on the hotplate (set it at 50 °C) for a few min. When the resin is soft, use a scalpel (with blade #11) to cut a square (~4 × 4 mm) of brain tissue

that contains the ROI. Remove the square with a wet tooth-pick, and glue it on the top of a resin block (~7 mm in diameter). Let the block dry at RT for 12 h and then cut the tissue on the ultramicrotome.

3.4 Ultramicrotomy (see Notes 26–31 and Fig. 2)

Transmission electron microcopy (TEM) usually requires that sections through biological specimens be less than 100 nm thick to allow electrons to pass. Thus, for 3D reconstruction and accurate quantitative morphometric analysis of specific neural elements, we must generate complete ribbons of serial ultrathin sections of perfectly uniform thickness. Because the morphometric data are often collected from immunostained elements, the sampling of structures to be reconstructed must be limited to the superficial part of the tissue (i.e., with optimal immunolabeling). The procedure for serial ultrathin sectioning is as follows:

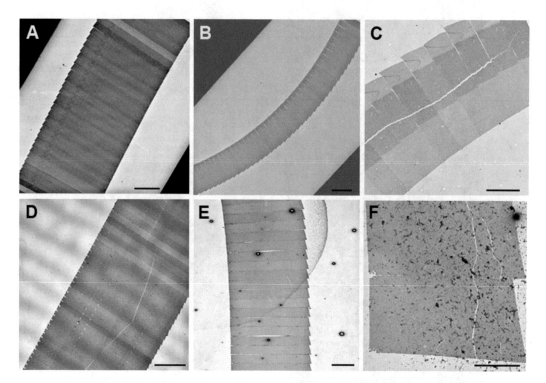

Fig. 2 Low-magnification images of transmission electron microscope grids showing examples of problems to overcome during ultramicrotomy and post-section staining. (**a**) Ribbon of serial uneven ultrathin sections; this can be due to different reasons, i.e., airdrafts, vibrations, heat variations, knife angle not parallel to block face, etc. . . . (**b**) Curved ribbon of sections is frequently the consequence of blocks trimmed with non-parallel top and bottom. (**c**) Ultrathin sections with tissue tracks, frequently due to a poor resin infiltration in the tissue. (**d**) Image of a thick and uneven pioloform membrane with a ribbon of uneven ultrathin sections. (**e**) Example of a pioloform membrane with holes caused by moisture. (**f**) Ultrathin sections covered with lead citrate precipitate; to avoid this problem keep lead citrate solution always in the fridge and do not use it if is older than a week. Scale bars (**a**)–(**f**) =100 μm. Images in this figure have been obtained from: http:/synapses.clm.utexas.edu/lab/howto/MeetExpert/MeetExpert.pdf

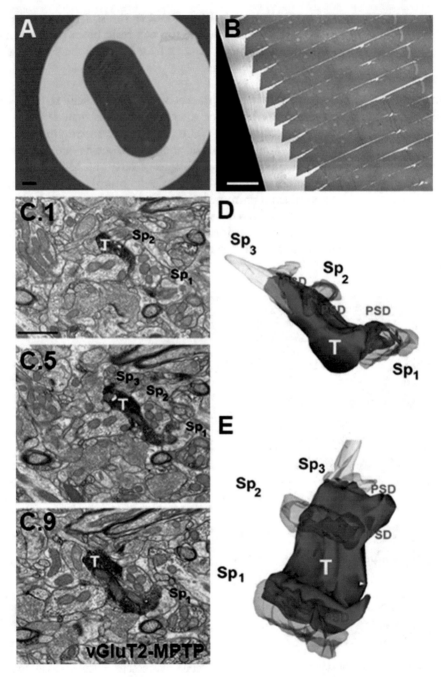

Fig. 3 (a) Light microscope image of an electron microscope pioloform-coated grid. (b) Low-magnification electron microscope image of a single slot grid with a ribbon of serial ultrathin (60–70 nm) sections. (c.1, c.5, c.9) Samples of serial electron micrographs used to generate the three-dimensional (3D) reconstructed images shown in (d). (d, e) 3D-reconstructed immunostained terminal (T) in contact with three different spines (Sp1, Sp2, Sp3). Note that the 3D-reconstructed spines are partially transparent and the image in (e) is rotated to better illustrate the postsynaptic densities. *D* dendrite, *Sp* spine, *PSD* postsynaptic density. Scale Bars = 200 μm in (a), 100 μm (b) and 1 μm in (c.1) (also applies to c.5, c.9). *Some images in this figure have been taken from: Villalba and Smith, 2011 JCN 519:989–1005*

1. Pioloform-coated slot grids are prepared as described in Section 2.7 (Fig. 3a).

2. The tissue block is trimmed into a trapezoid shape pyramid with top and bottom edges exactly parallel. It is critical that these two edges are parallel to produce long ribbons of ultrathin sections. This can be done:

   – Using the DiATOME cryotrim 45° knife. This knife allows creating very small trapezoids with smooth and parallel top and bottom edges, which facilitates serial thin sectioning. To ensure optimal use of this knife, the block must be tightly secured into the chuck of the ultramicrotome (Fig. 1b).

   – Placing the block in the trimmer stand. The trimmer stand is hollow so that the bottom light source can shine through the resin block. Using a new razor blade and looking down the binoculars, trim away the excess of resin from each of the four sides of the block in order to obtain a trapezoid-shaped pyramid. The four sides of the blocks must be carefully trimmed down again using another new razor blade so as to obtain the smallest possible block face that includes only the area of interest.

   The large width to height ratio of the trapezoid facilitates sections sticking together better as the ribbon is cut. A trapezoid, as small as 700–800 μm × 80–90 μm (Fig. 1c), can produce one long continuous ribbon of tightly linked sections (Fig. 1e).

3. When the block is ready, it has to be placed in the specimen holder and mounted onto the arm of the microtome set into the locked position.

4. The diamond knife DiATOME Ultrathin 45° is used to cut series of serial ultrathin sections (60–70 nm thick) (*see* **Note 24**).

5. When a sufficient number of sections are cut, the ultramicrotome is stopped, and the ribbon of sections is broken into shorter segments (small enough to fit onto Pioloform-coated slot grids) by tapping it with a fine eyelash near the juncture of two sections.

6. Usually, 15–25 serial sections, 60–70 nm thick (pale gold to silver) per grid (*see* **Note 25**) (Fig. 3a). The ribbon is collected by putting the grid under the liquid in the knife boat and then coming up under the sections with the grid.

3.5 Post-section Staining

1. After picking the sections, the grids need to be properly dried before staining with lead citrate.

2. For lead staining, it is recommended to use the grid staining matrix system (PELCO®) that is designed to handle from 1 up

to 25 grids at once, and has an alphanumeric system for identification of the grids. After placing the grids in the matrix, slide the cover over the body and place the matrix into an empty staining vessel. Add the appropriate volume of lead citrate (*see* Section 2.8). Stain grids for 5 min. Rinse the matrix with ddH$_2$O (rinsing of the grids can be done in a beaker or similar container) (*see* **Note 32**).

3.6 Serial Section Imaging

1. Two TEM are used in our laboratory: Zeiss EM-10C and Jeol electron microscopes with a CCD camera (GATAN-DualView 300 W).

2. For each grid, a low-magnification picture is taken of the entire grid to check the orientation of the sections and to count the number of sections on the grid (Fig. 3b).

3. Before collecting data from specific elements, warm up the pioloform coating and tissue at low magnification (50×) (*see* **Note 33**).

4. The ROI in middle sections of a series is then selected for imaging through serial sections (*see* **Notes 34** and **35**).

5. Once selected, the ROI is photographed at 16,000× magnification with a CCD camera (DualView 300 W; DigitalMicrograph software, version 3.10.1; Gatan, Pleasanton, CA) (Figs. 3c.1, c.5, c.9, and 4a, b).

3.7 Three-Dimensional Reconstruction

At this step, using ultrathin serial sections and images from the TEM, a specific structure or neuronal complex can be reconstructed (Fig. 3d, e).

1. Gatan EM images must be converted to new TIF files (*see* **Note 34**).

2. For the next steps, the RECONSTRUCT software must be downloaded from: http://synapses.clm.utexas.edu/ [6, 7, 15, 19, 22].

3. The series of images (TIFFs) are imported (*see* **Note 36**), and the section thickness adjusted (in our case to 60 nm) (*see* **Note 37**).

4. Alignment of the images: In this step, we align the whole field, rather than just individual contours or objects, ensuring that they are not artificially straightened or distorted (*see* **Note 38**).

5. To evaluate the quality of alignment on screen, section images are compared by blending and flickering. Blended images become sharper as they align. When all sections are satisfactorily aligned, the changes must be saved ("locked").

6. Tracing, analysis, 3D reconstruction, and image rendering: When the structures to be reconstructed are identified in all sections, the contour of their boundaries is generated by

manual tracing. By defining the contours of the objects on each section, the shape of the selected objects is defined from which the program generates a 3D representation and provides information about their dimensions (i.e., volume, surface, length, etc....). Then, color and transparency can be added to the reconstructed structures for final image production. This procedure results in final images of specific elements in 3D from which the morphology and morphometric measurements (e.g., PSD and axon terminal, dendritic spines, and spine apparatus) can be assessed (Fig. 4a.1, a.2, b.1, b.2, c)

3.8 Other Methods   Because 3D reconstructions using TEM are extremely time consuming and technically demanding, several new EM imaging methods have emerged, and classical EM methods have been automated to reduce the technical burden of serial ultrathin sectioning [32, 33]. With these new methods, not discussed in this chapter, serial images are obtained using scanning electron microscopy (SEM) with the detection of back-scattered electrons from the block face that is serially removed either by focused ion beams (FIB-SEM) [34, 35], or by a diamond knife inside the SEM chamber [36]. FIB-SEM combines the advantage of the SEM with the ion beam's ability to polish (mill) away a few nanometers at the time from the block face. Repeating this cycle produces a stack of sequential images representing 3D ultrastructures [37–40]. The FIB-SEM is limited by the size of the field that can be milled and imaged, usually less than 100mx100m [41]. Other method developed for imaging serial sections is using a scanning transmission electron microscopy (STEM) based on a field emission SEM fitted with a multi-mode transmitted electron detector [41, 42]. This system allows to reconstruct a large-volume of brain tissue using semi-automated acquisition of large-field images at high resolution from multiple grids (spanning an entire series of 200 serial sections) [41, 42]. The images taken with this system have less optical and physical distortions, and the resolution is comparable to those taken with a conventional TEM [41, 42].

4   Notes

1. The Ringer's solution can be prepared in advance, but needs to be stored in the fridge (4 °C), and kept on ice during perfusion.

2. Use ddH$_2$O to dissolve the paraformaldehyde and add later the PB to complete 2 l of fixative. Paraformaldehyde is a purified crystalline trimer of formaldehyde and under heat goes into aqueous solution without formic acid formation and electron-opaque artifacts [43].

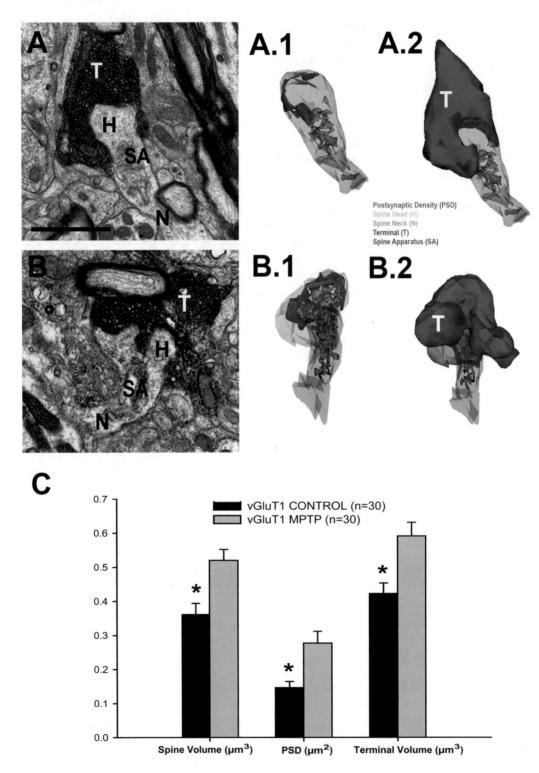

Fig. 4 Electron micrographs, three-dimensional (3D) reconstruction, and quantitative ultrastructural analysis of axo-spinous synapses in the monkey striatum. (**a, b**) Electron micrographs of synapses between the spine of a medium spiny neuron (MSN) and a vGluT1-positive terminal (T) in the striatum of a control (**a**)

3. Formaldehyde and glutaraldehyde are the most commonly used fixatives. The combination of formaldehyde with glutaraldehyde (Karnovsky's fixative) as a fixative for electron microscopy takes advantage of the rapid penetration of small aldehyde molecules, which initiate the structural stabilization of the tissue. The relative concentrations of the aldehydes can be adjusted based on the specific needs of the experiments.

4. Fixative solutions containing paraformaldehyde only must be freshly prepared 1–5 days before perfusion and kept in the fridge at 4 °C. Additional fixative, like glutaraldehyde or acrolein, must be added to the paraformaldehyde solution just prior to the perfusion to avoid denaturation.

5. Always use EM grade glutaraldehyde, which is generally supplied as 25 % aqueous solution from various EM material suppliers. The glutaraldehyde solution to be used must contain the monomer and low polymers (oligomers) with molecules small enough to penetrate the tissue fairly quickly. Commercially available non-EM grade glutaraldehyde solutions contain large polymers which are too large to fit between the macromolecules of cells and other tissue components.

6. The glutaraldehyde solution must be added to the paraformaldehyde only at the time of perfusion to avoid denaturation. The paraformaldehyde solution can be prepared in advance and stored at 4 °C.

7. DAB is carcinogenic. Always wear gloves and appropriate protective clothing. All material used in the reaction must be put in a bleach solution (1 part bleach in 9 parts of water) for at least 1 h to inactivate the DAB.

8. Special caution needs to be used when working with OsO_4. Because of the high toxicity of OsO_4, all experimental steps using this fixative must be done under the hood. Always wear gloves. The osmium waste must be placed in a disposal bottle that has to be kept always under the hood.

Fig. 4 (continued) and an MPTP-treated (**b**) monkey. (a.1, a.2, b.1, b.2) 3D-reconstructed images of the corresponding glutamatergic axo-spinous synapses depicted in (**a**) and (**b**) from the striatum of control (a.1, a.2) and MPTP-treated (b.1, b.2) monkeys. The head (H) and neck (N) of the spines are partially transparent to better show and compare the ultrastructural complexity of the postsynaptic density (PSD) and spine apparatus (SA). (**c**) Histograms comparing the morphometric measurements (spine volume, PSD area, terminal volume) of structural elements at corticostriatal (vGluT1-positive) glutamatergic synapses using the 3D reconstruction method of serial ultrathin sections collected from 30 axo-spinous synapses in normal and MPTP-treated monkeys. The spine volumes, the PSD areas, and the volume of vGluT1-containing terminals are significantly larger in MPTP-treated parkinsonian monkeys than in controls (*$P < 0.001$; t-test). *Some of the images in this figure have been taken from: Villalba and Smith, 2011 JCN 519:989–1005*

9. It is important to match resin and tissue hardness. Hard tissue embedded in a soft resin would break out of the resin block. Soft tissue in a hard resin would be distorted or disintegrate as it is moved across the cutting edge. Hardness of epoxy resin can be adjusted by changing the amounts of monomer components. The right amounts and the details to use it for different tissues are described in the manufacturer's instructions.

10. Use a room with a dehumidifier to coat grids. High humidity produces uneven and low quality Pioloform membranes.

11. The only way to determine if the thickness of the membrane is correct is to spread it over a water bath, and look at their color. When floating on the water surface, the membranes should display a gray tint with yellow strips. If they are too yellow, the solution is too concentrated.

12. Before using the grids to mount ultrathin sections, check the quality of the Pioloform membrane under the EM. The quality of this membrane has to be optimized in order to produce long lasting films with virtually no drift once in the column and stable under the electron beam. The membrane must have a uniform thickness and be without holes.

13. Lead citrate: All glassware for lead citrate preparation must be as clean as possible to prevent precipitation. Clean carefully each piece before use with hot soapy water and then rinse thoroughly. Lead citrate is toxic; always wear gloves to prepare this solution.

14. Vibratome sections in antifreeze solution can be stored in the freezer (-20 to -80 °C) for many years without alteration of their ultrastructural details.

15. The free aldehyde groups introduced by glutaraldehyde fixation cause various problems, including nonspecific binding of antibodies. These free aldehydes must be removed or blocked by appropriate histochemical procedures (i.e., incubation in sodium borohydride solution) before immunohistochemistry.

16. EM post-fixation mostly involves the use of osmium tetroxide (OsO_4). OsO_4 stabilizes cellular proteins, which allows preserving membrane components, and provides good image contrast.

17. OsO_4 penetrates tissue very slowly due to the low diffusion rate of its large molecules. The optimal time for tissue fixation in OsO_4 is normally 30–90 min. If the tissue is not fixed long enough, there may be inadequate stabilization of the specimen, whereas fixation for prolonged periods can occasionally lead to solubilization of some tissue components.

18. Uranyl acetate is a low level radioactive material. Caution must be taken while manipulating it. Wear lab coat and double

gloves. Use disposable plastic beakers to prepare the solution. Place all the contaminated glassware into the appropriate liquid waste container.

19. Most epoxy resins are immiscible with water and complete dehydration must be carried out prior to the use of epoxy. Any water in the tissue will result in a poor embedment. It's very important to use a transitional solvent, as propylene oxide, to aid in the infiltration of epoxy resins into the tissue.

20. Take great care when working with Durcupan ACM. Do not breathe in the vapor and avoid skin contact, because this may cause skin irritation and allergic reactions. Splashes on the skin must be washed off immediately with a 3 % boric acid solution. Frequent washing of hands, arms, and face with lukewarm soap water is advisable.

21. All the material that has been in contact with the resin (aluminum dishes, beakers, and dirty aluminum foil) must be put in the 60 °C oven for 48 h before being discarded. Brushes and forceps must be cleaned with propylene oxide.

22. The heat yields a strong 3D polymer which is resistant to solvents and heat. The 3D polymerization provides excellent mechanical strength for ultrathin sectioning.

23. It is very important to wear protective glasses during this procedure to avoid eye injuries.

24. Before and after sectioning, clean carefully the diamond knife. Bevel a styrofoam rod to an angle of approximately 60°; soak the beveled tip of the rod in ethanol 50 % and shake off the excess and then lightly draw it across the knife-edge without applying lateral pressure. Place the knife-edge under running distilled water to rinse any ethanol from it.

25. The section thickness is determined by the interference colors of the sections as they reflect light. Gray-silver sections range from 50 to 60 nm in thickness, and silver-gold sections range from 60 to 90 nm. Sections showing interference colors such as deep red, purple, pink, blue, green, brown, red, and violet are too thick for electron microscopy.

26. Difficulty wetting the knife-edge: Fill the boat with liquid (distilled water) until the water level is a little too high, and after a few minutes carefully remove the excess water.

27. Wetting the block face: For epoxy resins, the main cause of this problem is either overfilling the boat (causing the liquid to be dragged over the knife-edge by capillary action) or electrostatic charging. Solutions: Lower the water level slightly until there is an even silver reflection over the whole liquid surface or at least

around part of the knife-edge the sections are cut from. Dry the block face with filter paper.

28. If thin gold to silver sections are difficult to cut, or if sections are not being cut every stroke of the cutting cycle, stop the automatic cycling control and check the following conditions:

    (a) Cutting speed

    (b) Section thickness control

    (c) Proper illumination on microtome set to give white reflection on water surface

    (d) Floatation fluid in contact with the cutting edge of the knife

    (e) Floatation fluid not leaking from reservoir

    (f) Microtome controls sufficiently holding specimen and knife firmly

29. Chatter: the most common reasons that cause chatter are:

    (a) Screws are not full tightened (block, block holder, and knife)

    (b) Clearance angle is too small (may cause friction between the block face and the diamond face)

    (c) Cutting pressure too great

    (d) External vibrations

    (e) A faulty microtome

    Solutions include:

    (a) Make sure all screws are tightened.

    (b) Increase the clearance angle by 1–2° (usually cut a clearance angle 6°).

    (c) Reduce the block size and/or re-cutting another specimen.

    (d) Change the location of the microtome.

30. Sections compression: This occurs normally due to the physical process of section cutting and can be rectified by using chloroform or acetone. The main causes and possible solutions are:

    (a) The block is too soft: Try putting it back in the oven for a few days.

    (b) The knife is dull: Time to change the knife.

    (c) The clearance angle is too big and/or the cutting speed is too high: Reduce the clearance angle by 1–2° and/or reduce the cutting speed from 1 to 0.5 mm/s.

31. Nicks, tears, etc. to sections: Damage to the cutting edge of the knife will cause damage ranging from very fine lines through to tears in your sections.

The causes of knife-edge damage include:

(a) Remnant particles in your block from trimming

(b) Hard particles in your block and specimen.

(c) Normal knife use will cause nicks: During normal use, the knife-edge becomes less sharp.

Solutions: Use a new, clean razor blade for the final trimming of the block. This will prevent small, hard bits from attaching to the leading edge of the block and/or damaging the knife-edge.

32. During the lead staining, if air is trapped within a grid containing well it is easy to remove by moving the matrix to the front of the staining vessel and then quickly to the back. This will pull the trapped bubbles out of the wells. Be very careful not to spill reagents during this process.

33. Ultrathin sections supported on Pioloform-coated slot grids can be stabilized in the column of the TEM by exposing the area of interest to irradiation using low or intermediate magnifications. With this procedure we minimize the risk of tearing the Pioloform membrane in the vacuum of the TEM, the drift or distortion of the specimens during subsequent photography.

34. ROI from the middle of section series is chosen to make sure there are enough sections before and after the chosen sections to completely reconstruct the neuronal complexes under study (i.e., from a series of 24 sections/grid, the first sections to look at for ROI would be sections numbered 12–13).

35. To be able to recognize easily ROI (usually axo-spinous or axo-dendritic synaptic complexes) in all sections of a specific series, it is important to identify them in relation to some structures of reference (e.g., blood vessels, myelin tracts) that can easily be followed through many serial sections. Thus, although small elements in the vicinity of the ROI may disappear through analysis of the series of sections, the identification of these landmarks ensures that the same region is followed through all sections.

36. It is very important to keep the numerical order for the series during this conversion.

37. The section thickness is calibrated using the cylindrical mitochondrial method. This method uses the ratio of the maximum diameter of longitudinally sectioned mitochondria (or other cylindrical objects) to the number of serial sections they span [44].

38. Serial sectioning and imaging each section separately induces misalignment between sections. Also, aligning the whole field often allows several objects to be reconstructed and analyzed at once. Alignment of images also facilitates identification and tracing of objects. Difficulties in identifying a structure on a

single section can be resolved by following the structure on adjacent sections. This is much easier to do with aligned section images.

References

1. Ramon y Cajal S (1891) Sur la structure de l'ecorce cerebrale de quelques mammiferes. La Cellule 7

2. Palay SL, Palade GE (1955) The fine structure of neurons. J Biophys Biochem Cytol 1:69–88

3. Gray EG (1959) Electron microscopy of synaptic contacts on dendrite spines of the cerebral cortex. Nature 183:1592–1593

4. Spacek J (1987) Ultrastructural pathology of dendritic spines in epitumorous human cerebral cortex. Acta Neuropathol 73:77–85

5. Stevens JK, Trogadis J (1986) Reconstructive three-dimensional electron microscopy. A routine biologic tool. Anal Quant Cytol Histol 8:19102–19107

6. Harris KM, Stevens JK (1988) Dendritic spines of rat cerebellar Purkinje cells: serial electron microscopy with reference to their biophysical characteristics. J Neurosci 8:4455–4469

7. Harris KM, Stevens JK (1989) Dendritic spines of CA 1 pyramidal cells in the rat hippocampus: serial electron microscopy with reference to their biophysical characteristics. J Neurosci 9:2982–2997

8. Carlbom I, Terzopoulos D, Harris KM (1994) Computer-assisted registration, segmentation, and 3D reconstruction from images of neuronal tissue sections. IEEE Trans Med Imaging 13:351–362

9. Fiala JC, Spacek J, Harris KM (2002) Dendritic spine pathology: cause or consequence of neurological disorders? Brain Res Rev 39:29–54

10. Briggman KL, Denk W (2006) Towards neural circuit reconstruction with volume electron microscopy techniques. Curr Opin Neurobiol 16:562–570

11. Mishchenko Y (2009) Automation of 3D reconstruction of neural tissue from large volume of conventional serial section transmission electron micrographs. J Neurosci Methods 176:276–289

12. Ingham CA, Hood SH, Arbuthnott GW (1989) Spine density on neostriatal neurons changes with 6-hydroxydopamine lesions and with age. Brain Res 503:334–338

13. Ingham CA, Hood SH, Tagart P, Arbuthnott GW (1998) Plasticity of synapses in the rat neostriatum after unilateral lesion of the nigrostriatal dopaminergic pathway. J Neurosci 18:4732–4743

14. Harris KM, Jensen FE, Tsao B (1992) Three-dimensional structure of dendritic spines and synapses in rat hippocampus (CA1) at postnatal day 15 and adult ages: implications for the maturation of synaptic physiology and long-term potentiation. J Neurosci 12:2685–2705

15. Harris KM, Kater SB (1994) Dendritic spines: cellular specializations imparting both stability and flexibility to synaptic function. Annu Rev Neurosci 17:341–371

16. Picconi B, Pisani A, Barone I, Bonsi P, Centonze D, Bernardi G, Calabresi P (2005) Pathological synaptic plasticity in the striatum: implications for Parkinson's disease. Neurotoxicology 26:779–783

17. Stephens B, Mueller AJ, Shering AF, Hood SH, Taggart P, Arbuthnott GW, Bell JE, Kilford L, Kingsbury AE, Daniel SE, Ingham CA (2005) Evidence of a breakdown of corticostriatal connections in Parkinson's disease. Neuroscience 132:741–754

18. Zaja-Milatovic S, Milatovic D, Schantz AM, Zhang J, Montine KS, Samii A, Deutch AY, Montine TJ (2005) Dendritic degeneration in neostriatal medium spiny neurons in Parkinson disease. Neurology 64:545–547

19. Arellano JI, Espinosa A, Fairen A, Yuste R, DeFelipe J (2007) Non-synaptic dendritic spines in neocortex. Neuroscience 145:464–469

20. Deutch AY, Colbran RJ, Winder DJ (2007) Striatal plasticity and medium spiny neuron dendritic remodeling in parkinsonism. Parkinsonism Relat Disord 13:S251–S268

21. Surmeier DJ, Ding J, Day M, Wang Z, Shen W (2007) D1 and D2 dopamine-receptor modulation of striatal glutamatergic signaling in striatal medium spiny neurons. Trends Neurosci 30:228–235

22. Bourne JN, Harris KM (2008) Balancing structure and function at hippocampal dendritic spines. Annu Rev Neurosci 31:47–67

23. Bourne JN, Harris KM (2011) Coordination of size and number of excitatory and inhibitory synapses results in a balanced structural plasticity along mature hippocampal CA1 dendrites during LTP. Hippocampus 21:354–373

24. Bourne JN, Chirillo MA, Harris KM (2013) Presynaptic ultrastructural plasticity along CA3 → CA1 axons during long-term potentiation in mature hippocampus. J Comp Neurol 521:3898–3912

25. Smith Y, Villalba R (2008) Striatal and extra-striatal dopamine in the basal ganglia: an overview of its anatomical organization in normal and Parkinsonian brains. Mov Disord 23: S534–S547

26. Villalba RM, Lee H, Smith Y (2009) Dopaminergic denervation and spine loss in the striatum of MPTP-treated monkeys. Exp Neurol 215:220–227

27. Villalba RM, Smith Y (2010) Striatal spine plasticity in Parkinson's disease. Front Neuroanat 4:1–7

28. Villalba RM, Smith Y (2011) Neuroglial plasticity at striatal glutamatergic synapses in Parkinson's disease. Front Syst Neurosci 5:1–9

29. Villalba RM, Smith Y (2013) Differential striatal spine pathology in Parkinson's disease and cocaine addiction: a key role of dopamine? Neuroscience 251:2–20

30. Harris KM, Weinberg RJ (2012) Ultrastructure of synapses in the mammalian brain. Cold Spring Harb Perspect Biol 4:1–30

31. Kuwajima M, Spacek J, Harris KM (2013) Beyond counts and shapes: studying pathology of dendritic spines in the context of the surrounding neuropil through serial section electron microscopy. Neuroscience 251:75–89

32. Briggman KL, Bock DD (2012) Volume electron microscopy for neuronal circuit reconstruction. Curr Opin Neurobiol 22:154–161

33. Denk W, Horstmann H (2011) Serial block-face scanning electron microscopy to reconstruct three-dimensional tissue nanostructure. PLoS Biol 2:e329

34. Langford RM (2006) Focused ion beams techniques for nanomaterials characterization. Microsc Res Tech 69:538–549

35. Knott G, Marchman H, Wall D, Lich B (2008) Serial section scanning electron microscopy of adult brain tissue using focused ion beam milling. J Neurosci 28:2959–2964

36. Smith SJ (2007) Circuit reconstruction tools today. Curr Opin Neurobiol 17:601–608

37. Helmstaedter M, Briggman KL, Denk W (2008) 3D structural imaging of the brain with photons and electrons. Curr Opin Neurobiol 18:633–641

38. Hayworth KJ, Xu CS, Lu Z, Knott G W, Fetter R, Tapia JC, Lichtman JW, Hess HF (2015) Ultrastructurally smooth thick partitioning and volume stitching for large-scale connectomics. Nat Methods 12:319–322.

39. Merchan-Perez A et al (2009) Counting synapses using FIB/SEM microscopy: a true revolution for ultrastructural volume reconstruction. Front Neuroanatal 3:18

40. Bosch C, Martinez A, Masachs N, Teixeira CM, Fernaud I, Ulloa F, Perez-Martinez E, Lois C, Comella JX, DeFelipe J, Merchan-Perez A, Soriano, E (2015) FIB/SEM technology and high-throughput 3D reconstruction of dendritic spines and synapses in GFP-labeled adult-generated neurons. Front Neuroanat 9, Article 60

41. Kuwajima M, Mendenhall JM, Harris KH (2013) Large-volume reconstruction of brain tissue from high-resolution serial section images acquired by SEM-based scanning transmission electron microscopy. Methods Mol Biol 950:253–273

42. Kuwajima M, Mendenhall JM, Lindsey LF, Harris KM (2013) Automated transmission-mode scanning electron microscopy (tSEM) for large volume analysis at nanoscale resolution. PLoS One 8:e59573

43. Hayat MA (1981) The production of artifacts. Ultrastruct Pathol 2:93

44. Fiala JC, Harris KM (2001) Cylindrical diameters method for calibrating section thickness in serial electron microscopy. J Microsc 202:468–472

Neuromethods (2016) 115: 105–123
DOI 10.1007/7657_2015_75
© Springer Science+Business Media New York 2015
Published online: 07 January 2016

Electron Microscopy of the Brains of *Drosophila* Models of Alzheimer's Diseases

Kanae Ando, Stephen Hearn, Emiko Suzuki, Akiko Maruko-Otake, Michiko Sekiya, and Koichi M. Iijima

Abstract

The fruit fly *Drosophila* is widely used as a genetic model organism and recently emerged as a powerful system in which to study human diseases. We established fly models of Alzheimer's disease (AD) by expressing AD-associated β-amyloid peptides or microtubule-associated protein tau in the fly brain. Electron microscopy (EM) is an essential tool used to diagnose and categorize human diseases and to evaluate whether transgenic models recapitulate pathological phenotypes. We employed EM analyses to gain an understanding of the pathological effects of expressing Aβ or tau on the ultrastructure of the brain and to localize β-amyloid within subcellular organelles. These analyses revealed that several critical pathologies observed in the brains of patients with AD are recapitulated in these fly models of the disease.

Keywords *Drosophila*, Neurodegeneration, Alzheimer's disease, β-Amyloid, Tau, Electron microscopy, Immunogold labeling, Antigen retrieval, Osmium tetroxide, Mushroom body, Photoreceptor neuron

1 Background

The fruit fly *Drosophila* has a long history as a model organism in basic biology fields, including genetics, development, cell biology, neuroscience, and aging. Genomic data revealed that important biochemical and developmental pathways are conserved between fruit flies and humans, and approximately 70 % of human disease-related genes have homologs in *Drosophila* [1, 2]. *Drosophila* is now widely utilized in medical research fields to model human diseases in order to investigate the molecular mechanisms underlying disease pathogenesis. Furthermore, *Drosophila* serves as a platform for pharmacological screens designed to search for potential therapies.

The pathogenesis of many human diseases is complex because multiple genetic and environmental factors are involved. Elucidating diseases occurring in the central nervous system, such as neurodegenerative diseases, is especially challenging because the brain has complex wiring with heterogeneous populations of cells and because dysfunctions in and miscommunications between these different cell types contribute to disease pathogenesis. With its controlled genetic background and accumulated genetic resources, *Drosophila* provides advantages of genome-wide analyses for distinguishing causative factors and consequences in vivo in a cost- and time-efficient manner. Moreover, although it is much simpler than that in the human, *Drosophila* has well-organized brain wiring that mediates higher order brain functions, such as learning and memory. During the past decade, many neurological and neurodegenerative diseases have been successfully modeled in *Drosophila* [3–5].

Alzheimer's disease (AD) is a fatal disorder. The disease progression starts with short-term memory impairments presenting along with other psychiatric problems, such as sleep disorders and increased agitation. In the later stages, global cognitive functions are disrupted, and the associated motor disabilities lead patients to become bedridden [6, 7]. At the level of cellular pathology, extensive neuronal loss and two characteristic hallmarks, senile plaques (SPs) and neurofibrillary tangles (NFTs), are observed in the brains of patients with AD [8]. SPs are extracellularly deposited protein aggregates of which the major components are the 38–43 amino acid amyloid-β (Aβ) peptides [9, 10]. Although a small number of SPs are detected in normal-aged brains, this lesion is relatively specific to AD. By contrast, NFTs, which are intracellular protein inclusions composed of the hyperphosphorylated microtubule-associated protein tau [11], are observed in a class of neurological diseases called tauopathies that include AD. Mounting evidence from biochemical and genetic studies as well as findings from cellular and animal models suggest that the accumulation of Aβ in the brain, which starts decades before the onset of disease, leads to a number of pathologies, including tau abnormalities and neurodegeneration [12].

Transgenic *Drosophila* overexpressing human Aβ peptides or tau in the central nervous system have been used in many laboratories to investigate mechanisms underlying the neurodegeneration in AD [13, 14]. Fly models of Aβ toxicity have been established by others and us by expressing human Aβ peptides in the secretory pathways of neurons in the fly brain [13, 15]. The 42 amino acid Aβ peptide (Aβ42) expressed in fly brain neurons accumulates during aging and causes age-dependent deficits in learning, memory, and locomotion, as well as neuronal loss and premature death [13]. Fly models of human tau toxicity have been created by expressing the human wild-type or mutant forms of tau that are related to frontotemporal dementia with parkinsonism linked to chromosome 17 (FTDP-17) [16, 17]. Expression of human wild-type or mutant tau in neurons causes neuronal dysfunction, behavioral deficits, neuronal death, and reduced life-span [16–20].

An animal model of a human disease is expected to recapitulate at least one of the pathological hallmarks observed in human tissues. Electron microscopy (EM) analysis is an essential tool not only to diagnose and categorize human diseases at an ultrastructural level, but also to evaluate whether pathological phenotypes, such as cellular damage or protein aggregation, are recapitulated in an animal model to demonstrate the feasibility of using that animal model to investigate human disease. Moreover, EM analysis provides critical information about how genetic and pharmacological manipulations affect cellular phenotypes, which is critical for understanding mechanisms underlying the disease pathogenesis. In this chapter, using *Drosophila* models of AD as an example, we describe applications of two widely used EM techniques, namely transmission electron microscopy (TEM) and immunoelectron microscopy, for the analysis of fly brains [13, 14, 18].

2 Analysis of Neurodegeneration Caused by Human Aβ or Tau in the Brains of Transgenic *Drosophila* Using TEM

2.1 Introduction

Drosophila has several features that are advantageous for TEM analyses. First, its brain is small, less than 0.5 mm in length (Fig. 1). Since the entire brain can fit in a single thin section for TEM, it can be wholly examined to find the regions of interest. Correlated light and electron microscopy (CLEM) can be applied to the whole brain. The small brain size also simplifies fixation processes; fly brains do not require perfusion but can be fixed by immersion in a fixative solution. Second, comprehensive atlases and databases of the *Drosophila* nervous system have been established

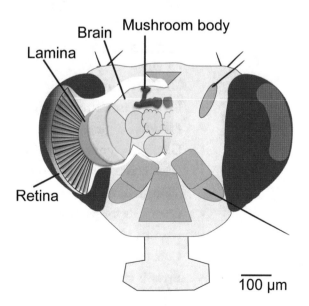

Fig. 1 Schematic representation of the fly brain. The mushroom body (*red*) and the lamina (*green*) are indicated

and are freely available [21]. These resources can be used to identify the regions of interest in the thick sections for TEM sample blocks.

Several major differences between mammals and *Drosophila* at cellular and tissue levels should be considered when using *Drosophila* as a model system for human disease. First, flies do not have a cardiovascular system but a tracheal system that brings air directly to the organs. Second, the locations of the neurons in the fly brain are different from those in the mammal brain. Third, *Drosophila* cells lack intermediate filaments.

A tissue-specific expression system, the GAL4-UAS system [22], was used to express Aβ or tau in *Drosophila* models of AD. Aβ was expressed in all neurons using a pan-neuronal driver, and TEM analyses were conducted, focusing on the mushroom body, a learning and memory center in the fly brain (Fig. 1). Tau was expressed in the eyes of *Drosophila* using a pan-retinal driver, and to observe photoreceptor axons, TEM analyses were focused on the lamina (Fig. 1), the first synaptic neuropil of the optic lobe containing photoreceptor axons.

2.2 Materials

2.2.1 Materials and Equipment

- Forceps No. 5 antiacid, antimagnetic type, one for dissection and the other for handling grids.
- Razor blades.
- Silicon rubber plate or dental wax plate as a base plate for dissection.
- 2 mL 100 % polypropylene microfuge tubes.
- 30–50 mL screw-capped 100 % polypropylene tube (PPT) with a scale of 1 mL division.
- Rotator.
- Embedding mold for sample embedding and for sample mounts.
- 50–70 °C incubator.
- Electric disc cutter or a jigsaw.
- Fast-drying glue.
- Ultramicrotome with a trimming stage.
- Glass knives and a diamond knife.
- Slide glasses.
- Glass rods for picking thick sections.
- Heating plate.
- Copper grids type 300HH or 300HS thin bar mesh.
- Eyelash cleaned with acetone thoroughly and mounted on a thin stick.
- Petri dishes with 20–30 mm diameter.
- Filter paper #1.

- Light microscope for checking thick sections.
- Transmission electron microscopes.

- Primary fixative; 2 % paraformaldehyde + 2.5 % glutaraldehyde in 0.1 mol/L sodium cacodylate buffer (pH 7.3).
  - 2 % paraformaldehyde is diluted from freshly made 20 % paraformaldehyde.
  - 20 % paraformaldehyde is made by heating the solution up to 60 °C and add minimum amount of 1 N sodium hydroxide to dissolve paraformaldehyde powder. This procedure should be done in a draft chamber.
  - 0.1 mol/L sodium cacodylate buffer is diluted from 0.2 mol/L stock solution.
  - 2.5 % glutaraldehyde is diluted from 25 % EM-grade glutaraldehyde, which is commercially available.
- Secondary fixative: 1 % OsO_4 in 0.1 mol/L sodium cacodylate buffer (pH 7.3).
  - Mix 1 volume of 4 % OsO_4, 1 volume of H_2O, and 2 volumes of 0.2 mol/L sodium cacodylate stock solution. OsO_4 is volatile and the vapor is toxic. This process should be done in a draft chamber.
- 0.5 % uranyl acetate in H_2O for block staining and 2 % uranyl acetate in 70 % ethanol for section staining.
- 50 %, 70 %, 90 %, 100 % ethanol (EtOH).
- Propylene oxide.
- Resin: Mix 10 mL of Epon 812, 7 mL of MNA, and 3 mL of DDSA for 3 min in a PPT, then add 0.4 volume of DMP-30, and mix well again for 20 min. Rotate with hands for the first 3 min, and then further rotate on a rotator for 20 min at a slow speed. After mixing, keep the tube still for 1 h to eliminate air bubbles.
- 1 % toluidine blue O in 1 % sodium borate solution.
- Reynolds' lead solution:
  - Dissolve 2.66 g lead nitrate $Pb(NO_3)_2$ in 30 mL H_2O.
  - Dissolve 3.52 g trisodium citrate dehydrate $Na_3(C_6H_5O_7) 2H_2O$ in 30 mL H_2O.
  - Add above solutions and stir for 30 min. The mixture becomes turbid white.
  - Add 16 mL 1 N NaOH and mix again until the solution becomes transparent.
  - Transfer the solution to 100 mL measuring cylinder.
  - Add H_2O to make the solution 100 mL and mix.

2.3   Methods

2.3.1   Fixation and Embedding

1. Anesthetize flies with CO_2 or with ether for 30 s to 1 min.

2. Decapitate the flies with a razor blade.

3. Transfer the heads into a drop of primary fixative on a wax or silicon rubber plate.

4. Remove the probosces with a forceps making openings as large as possible.

5. Transfer the heads into the primary fixative in a microfuge tube and shake the tube for 10 s in order to make the fixative penetrate into the heads through the opening (Section 2.4, **Note 2**).

6. Fix for several hours at room temperature.

7. Wash in the buffer containing 3 % sucrose 3 × 5 min. The steps hereafter are done by changing solution in microfuge tubes (Section 2.4, **Note 3**).

8. Fix in 1 % OsO_4 in 0.1 mol/L cacodylate buffer, for 1–2 h on ice.

9. Wash in H_2O 3 × 5 min on ice.

10. Stain in 0.5 % aqueous uranyl acetate for 1 h on ice.

11. Dehydrate through ethanol series: 5 min in 50, 70, 90 % EtOH on ice; 100 % EtOH 3 × 5 min at room temperature; 2 × 5 min propylene oxide at room temperature.

12. Infiltrate with a mixture of propylene oxide and resin (1:1) overnight at room temperature (Section 2.4, **Note 4**).

13. Open the tube containing samples and keep in a desiccator for 1–2 h to increase the concentration of resin by evaporation of propylene oxide.

14. Pour fresh resin in embedding mold and put samples in it.

15. Keep at room temperature until the samples sink on the bottom of the mold. This takes about 1 h.

16. Adjust the orientation of the samples so that the posterior side faces the bottom, and the opening of the head cuticle is close to the edge of the mold.

17. Fill in a mold for sample mounts with the remaining resin.

18. Cure the resin in an incubator at 70 °C for 3 days.

*2.3.2   Sectioning and Observation (Section 2.4, **Note 5**)*

Figure 1 shows schematic representation of the fly brain. For the examination of the photoreceptor axons in laminas, horizontal sectioning starting from the opening of the excised proboscis is convenient. For the examination of mushroom body calyx and cell bodies, frontal sectioning starting from the posterior side is convenient.

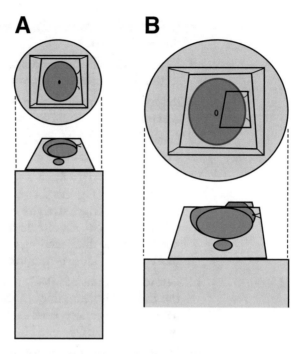

Fig. 2 An illustration of a sample block mounted for frontal sectioning of the mushroom body. (**a**) Top view (*upper panel*) and a side view (*lower panel*) of a sample block on a mount. Note that the posterior side of the head faces up. (**b**) Enlarged view of a re-trimmed sample block. A smaller trapezoid in the upper panel is the area for thin sectioning

1. Trim the resin block using electric disc cutter or a jigsaw, and fix the trimmed block on a sample mount with fast-drying glue so that the convenient side of the sample faces toward the top (Fig. 2).

2. Trim off the resin around the sample further with a clean razor blade using a trimming stage. Cut away resin to form a trapezoid shape, so that the final form resembles a pyramid with its top lopped (Fig. 2).

3. Cut thick sections of 1 μm thick with glass knives using an ultramicrotome until the appropriate plane of the tissue is exposed. Follow the manufacturer's instruction for how to make glass knives and cut sections.

4. Pick up the last thick section with a glass rod and transfer to a drop of water placed on a slide glass.

5. Warm up the slide glass on a heating plate (about 80 °C) and allow the water to evaporate.

6. After the water is completely dried, apply toluidine blue solution on the section and incubate for 30 s to 1 min.

7. Rinse the slide thoroughly with water, and dry up the water on the heating plate.

8. Examine the slide under a light microscope to see whether the section includes appropriate area.

9. Repeat sectioning and observation until the desired tissue area is exposed.

10. Select the area in the thick section for thin sectioning smaller than 0.5×0.5 mm^2.

11. Trim the block face further to form a trapezoid just around the selected area (Fig. 2b). Use a clean and sharp razor blade to make the cut edges straight and smooth. Top and bottom lines of the trapezoid should be at the right angle to the cutting motion of the thick section, so that readjustment of the sample block and the knife is minimized (Fig. 2b).

12. Cut thin sections of 50–70 nm thick using a diamond knife. If the top and bottom lines of the trapezoid are smooth enough to make the sections stick each other, they form a long ribbon on the water surface filled in the trough of the diamond knife.

13. Separate the section ribbon to ~2 mm long by gently tapping the side of a ribbon using an eyelash mounted on a thin stick.

14. Pick up the separated ribbons onto a grid. Hold a grid with No. 5 forceps at 45° against the water surface, go into the water underneath the ribbons to be picked up, and slowly lift the grid out of the water, while keep the ribbons on the grid using the eyelash.

15. Remove water on grids with a small piece of filter paper by gently touching the tip of holding forceps.

16. Dry the grids with sections on filter paper in a petri dish. Sections are made adhered to grid bars firmly by warming in a 50 °C incubator about 1 h to overnight.

17. Stain sections with 2 % uranyl acetate in 70 % ethanol for 5 min. The following steps are done in petri dishes. Transfer the grids from the staining dish to the first washing dish and so on.

18. Wash in H_2O three times.

19. Stain with Reynolds' lead solution for 5 min.

20. Wash in H_2O three times.

21. Remove water on grids with filter paper.

22. Dry grids with sections on filter paper in a petri dish.

23. Examine the sections in a transmission electron microscope at the acceleration voltage of 80 kV (Section 2.4, **Note 6**).

2.4 Notes

1. Most of the chemicals used for the fixation and staining of samples are toxic, so follow the manufacturer's safety guides when you use these materials.

2. Primary fixation is crucial for the preservation of tissue ultrastructure. The fly heads should be transferred into the fixative immediately after decapitation.

3. During fixation and embedding procedure, tissue should not be dried, so upon changing the solution, leave the top surface of previous solution above the samples, add the next solution, and mix.

4. Infiltration time of resin can be extended to several days.

5. Tools for sectioning should be clean. The forceps, eyelash, and grids should be cleaned with acetone to eliminate oily contamination. Copper grids should be cleaned in 1 N HCl for several seconds to remove CuO before acetone cleaning.

6. Upon examination of sections in an electron microscope, quality of fixation should be checked. Successful fixation and embedding give good preservation of membranous structures such as plasma membranes and intracellular organelles. Lipid bilayers of intracellular membranes sectioned at the right angle against the plane of the membrane appear as two parallel lines at high magnification. As for the neuropiles in the brain, neurites and synapses should be clearly identified by their ultrastructures, e.g., microtubules in axons and dendrites, and synaptic vesicles in synapses.

2.5 Typical Results

We used TEM to investigate how the cellular pathology observed in the brains of patients with AD is recapitulated in fly models of AD. AD brains are characterized by an extensive loss of neurons. At the ultrastructural level, degenerating neurons in an AD brain contain abnormal neurites filled with lysosomes and other laminated vesicles, granulovacuolar degeneration consisting of vacuoles with large granules of varying densities, paracrystalline structures called Hirano bodies, intranuclear rods, dystrophic axons, amyloid fibrils, and paired helical filaments [23].

Flies that express Aβ42 or tau recapitulate some of the degenerative features observed in AD brains [13, 14, 18]. Neurodegeneration is observed in the fly brain as a vacuolar appearance at light microscopy levels ([14], Fig. 3a, closed arrowheads). TEM analyses reveal that the majority of dying neurons in the brains of flies expressing Aβ42 show features typical of necrotic type cell death (Fig. 3b), that is, digested cytoplasm (electron-lucent) with swollen mitochondria (arrow) but relatively intact nuclei (indicated by N). Granulovacuolar degeneration is also observed (open arrowhead). Expression of human tau in the photoreceptor neurons causes age-dependent neurodegeneration, which is seen as vacuoles in the

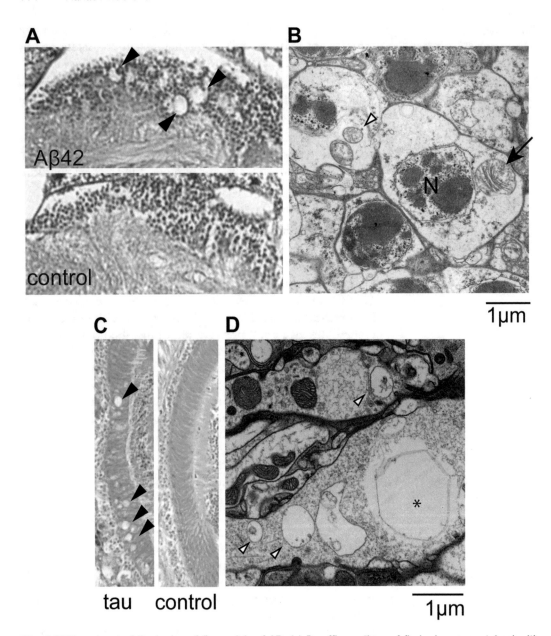

Fig. 3 TEM analyses of the brains of fly models of AD. (**a**) Paraffin sections of fly brains were stained with hematoxylin and eosin and observed at light microscopy levels. *Closed arrowheads* indicate neurodegeneration in flies expressing Aβ42 in the brain (*top*). Neurodegeneration was not observed in control brains (*bottom*). (**b**) Ultrastructural analysis of degenerating neurons in Aβ42 fly brain. Digested cytoplasm (electron lucent), swollen mitochondria (*arrows*), and vacuole containing a large granule of varying density (granulovacuolar degeneration, *open arrowheads*). *N* nucleus. (**c**) *Left*: Neurodegeneration (*closed arrowheads*) in the lamina in the flies expressing human tau. *Right*: The lamina of control flies. (**d**) Electron micrograph of the lamina area in a fly expressing human tau shows abnormal neurites containing circular profiles that were filled with laminated bodies (*open arrowheads*) and vacuole (*asterisk*)

lamina area in the brain at light microscopy levels (Fig. 3c, closed arrowheads). Electron micrographs of photoreceptor axons in flies expressing tau show swollen axons and vacuoles (Fig. 3d, asterisk). Abnormal neurites, consisting of circular profiles filled with lysosomes and other laminated bodies, are frequently observed (Fig. 3d, open arrowheads).

Extracellular deposits of amyloid plaques composed of Aβ and intracellular NFTs of the tau protein are the neuropathological hallmarks of AD in the human brain [8]. Despite extensive neurodegeneration, these structures are not observed in the brains of flies expressing Aβ or tau [13, 16]. These results suggest that neurodegeneration occurs without formation of fibrils, which is consistent with recent reports that prefibrillar forms of aggregation-prone proteins are major sources of neurotoxicity [24, 25].

3   Ultrastructural Localization of Human Aβ42 in Transgenic *Drosophila* Using a Post-embedding Immunogold Method on Samples Fixed with Osmium Tetroxide and Embedded in Plastic Resin

3.1   Introduction

Osmium tetroxide fixation of tissues prior to resin embedding was introduced by Georges Palade in 1952 to stain membranes in cells and tissues for examination under the electron microscope [26]. That protocol and its subsequent iterations were quickly adopted by neuroscientists, and its simplicity and wide use have made it the recognized gold standard protocol for ultrastructural imaging of neural tissues [27]. Osmicated and resin-embedded tissues were subsequently shown to be useful for post-embedding immunogold labeling after sections were treated with chemicals or heat to retrieve antigens [28–30]. This method was further extended for use on reduced osmium tetroxide-fixed samples embedded in the acrylic resin LR White, and surprisingly good results were achieved with membrane antigens, even where osmium was present. Neuroscientists interested in localizing Aβ42 infrequently use post-embedding gold labeling techniques on osmicated human brain tissues because osmium reportedly causes the loss of most antigenic epitopes, and with postmortem human brain tissue, autolysis is a further discouraging factor [31, 32]. Rapid cryofixation and freeze substitution without osmium have been used in resin-embedded tissues from animal models of Aβ42 to localize both extracellular and intracellular Aβ42 [33, 34]. Aβ antigenicity is preserved in extracellular amyloid fibril deposits of osmicated tissues [35]. In one report, Aβ could be localized to autophagic vacuoles in human brain tissue after osmication and resin embedding [36].

We developed transgenic fly lines that expressed human Aβ at robust levels as determined by light microscopic immunocytochemistry [13, 14]. We initially embedded these in LR White resin and Lowicryl K4M resin without osmium tetroxide post-fixation and found labeling consistent with that from the light microscope results; however, the absence of membranes limited the value of these EM results. Thus, post-osmicated brains were embedded in LR White resin, and thin sections received a brief treatment in hydrogen peroxide prior to immunogold labeling, giving strong and specific labeling for Aβ.

3.2 Materials

3.2.1 Reagents

- Phosphate-buffered saline (Difco FA buffer Becton, Dickinson, MA).
- 8 % glutaraldehyde (Electron Microscopy Sciences EMS, Hatfield, PA).
- Paraformaldehyde in Prills (Sigma Aldrich, St. Louis, MO).
- Bovine serum albumin (BSA-e Aurion EMS).
- Rabbit antibody to Abeta 1-42 AP5078P (Millipore, Temecula, CA).
- Goat anti-rabbit antibody conjugated to 10 nm colloidal gold (Aurion EMS).
- 10 % hydrogen peroxide, molecular biology-grade water (Sigma-Aldrich) (*see* Section 3.4, **Note 1**).
- 2 % uranyl acetate (EMS).
- 3 % potassium ferrocyanide $K_4FeCN_6 \cdot 3H_2O$ (Sigma Aldrich) in distilled water.
- 2 % osmium tetroxide in distilled water.
- Absolute ethanol.
- LR White resin medium-grade catalyst included (EMS).

3.2.2 Equipment

- Shaking rotator (Labline, Melrose Park, Il).
- Vented oven 60 °C.
- Transparent gelatin capsules 00 size, parafilm "M" laboratory film (Neenah WI), 100 mm square petri dish with grid (Thermo Scientific).
- 20 mL capacity clear glass scintillation vials with plastic lids (Thermo Scientific).
- Disposable transfer pipettes 1 mL capacity 100 mesh nickel grids (Veco EMS).
- Negative action anti-capillary antimagnetic Dumoxel with super thin tips Dumont N5AC forceps (EMS).
- Whatman #5 filter paper (VWR Scientific).
- Glass scintillation bottles 20 ml (VWR Scientific).
- BEEM capsule holder 00 size transparent (Ted Pella, Redding, CA).

3.3 Methods

3.3.1 Fixation

1. Probosces are removed from decapitated heads and primary fixation done overnight at room temperature by immersion in freshly prepared 4 % glutaraldehyde and 2 % paraformaldehyde in 0.1 mol/L PBS (*see* Section 3.4, **Note 2**).

2. Heads are transferred to 20 mL clear glass scintillation vials with plastic lids (Thermo Fisher Scientific), rinsed once in distilled water, and postfixed for 1 h in 1 % osmium tetroxide in 1.5 % potassium ferrocyanide in distilled water (*see* Section 3.4, **Note 3**).

3.3.2 Resin Embedding

1. Take LR White resin from the 4 °C refrigerator and add to an equal volume of 100 % ethanol and let it warm to room temp.

2. Osmium tetroxide fixative is removed from each vial and the brains rinsed in distilled water for 5 min and then dehydrated in a graded series of ethanol (30, 50, 100 %) with 5 min for each step. Infiltrate heads in 50 % resin for 1 h and then infiltrate overnight in 100 % resin. The samples are transferred to gelatin capsules filled with pure resin and polymerized overnight at 60 °C (*see* Section 3.4, **Note 4**).

3.3.3 Thin Sectioning

100 nm thick sections of mushroom body region are collected on 100 mesh nickel grids with each grid held in self-locking forceps and dried thoroughly at 37 °C for 10 min or longer by placing the forceps in an oven (*see* Section 3.4, **Note 5**).

3.3.4 Immunogold Labeling and Counterstaining

1. Prepare a primary incubation chamber and two rinse chambers from petri dishes with a square of hydrophobic parafilm attached to the bottom of the dish and a water-soaked filter paper to prevent evaporation of reagents (Fig. 4a). Dispense drops of each solution in vertical rows starting from left with 10 % hydrogen peroxide, 1 % BSA in PBS, primary antibody diluted 1:10 in PBS, and secondary antibody diluted 1:10 in PBS.

2. Make a rinse incubation chamber with rows of five drops of PBS. Make a second rinse incubation chamber with rows of ten drops of water.

3. Incubate grids with attached sections in H_2O_2 for 2 min for antigen retrieval, and jet rinse the section with water (ten drops from transfer pipette) (*see* Section 3.4, **Note 6**).

4. Place grids in BSA for 5 min.

5. Transfer grids into drops of primary antibody for 2 h at room temp.

6. Remove excess unbound primary from the grids by moving them through five drops of PBS (5 s each drop).

7. Place grids in secondary antibody-gold conjugate and incubate for 30 min. Rinse off unbound secondary antibody by rinsing in drops of water and allow sections to dry thoroughly.

8. Counterstain for 5 min with 3 % uranyl acetate dissolved in 30 % ethanol and rinse in distilled water.

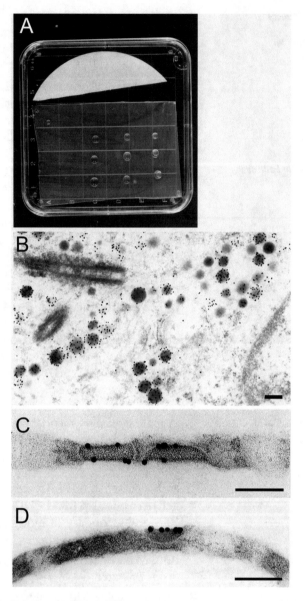

Fig. 4 Immersion of thin sections for improved immunogold labeling. (**a**) Incubation chamber with nickel grids bearing sections is immersed in drops of immune solutions dispensed on sheet of parafilm in petri dish with strip of moistened filter paper to prevent evaporation. (**b**) Thin section of human tumor (medullary carcinoma of the thyroid) viewed in the usual orientation with the grid at 90° to the electron beam showing positive labeling for calcitonin over neuroendocrine granules. (**c** and **d**) Side on view of thin section from sample shown in (**b**) after re-embedding and thin sectioning that indicates the labeling occurs only on the surface of resin sections and demonstrates the value of section immersion for maximizing the efficiency of labeling. All scale bars equal 100 nm

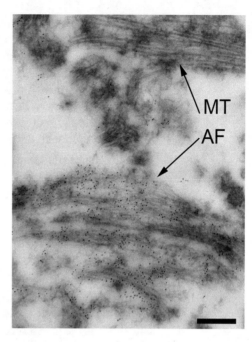

Fig. 5 Positive control for Aβ immuno-EM is a section of postmortem human brain from the frontal lobe region of the brain of an AD patient. Specific labeling is present over amyloid fibrils (AF) while nearby microtubules (MT) are negative. Autolysis is evident in this sample. Scale bar equals 500 nm

3.3.5 Immunogold Labeling Controls

1. For an ideal positive control obtain fixed human tissue from a brain bank from an AD patient (Fig. 5).

2. For negative controls incubate sections in drops containing either no primary antibody or an irrelevant antibody and continue with incubation steps in secondary antibody gold. Antigen retrieval results in a noticeable reduction in the density of membrane contrast due to the partial removal of osmium stain and the duration of time in which sections are exposed to antigen retrieval may need to be reduced to get ideal results.

3.4 Notes

1. All aqueous solutions for critical steps are made with molecular biology-grade water—especially those involving thin sectioning (i.e., the knife water bath solution), immunogold labeling, and heavy metal staining. This grade of water dramatically reduces the contamination of sections with precipitates of various kinds.

2. Primary fixative solution is used at room temperature to preserve microtubules and the fixation is done in 1.5 mL microcentrifuge tubes.

3. Solution changes are done quickly, using disposable pipettes being careful not to allow the brains to dry out, also being

careful not to dispose of the heads accidentally in waste solution. This and subsequent steps prior to embedding were done with agitation on a shaking rotator placed inside a chemical fume hood. Scintillation vials make it easy to see the heads, and manipulate them with pipettes, and they will not tip over with the rocking agitation. Potassium ferrocyanide-reduced osmium turns the brains a light brown color and specifically stains membranes and glycogen particles—always handle in fume hood.

4. Each head was transferred to the large half of a gelatin capsule containing a paper label curled to the top of the capsule with the sample ID done with a lead pencil. The capsule halves were placed on a BEEM capsule support rack which has holes that will support the capsules and keep them in an upright position. Resin was added to each capsule and the brains positioned with the eyes facing up the top half applied to make an airtight seal as oxygen prevents polymerization.

5. Drying prevents the sections from falling off. Generally in a single run about eight grids are a reasonable number to collect. 100 mesh grids give a nice large viewing area but one needs to have the largest possible section to have multiple attachment points to minimize section falloff. Knife quality is important.

6. Hydrogen peroxide is highly caustic and wear gloves and protective eyewear to guard against splashes. It is important to immerse the grids in each immune reagent to get labeling on both sides (Fig. 4b, c). Use of anti-capillary forceps prevents carryover of reagents.

3.5 Typical Results

Transgenic fly models of human AD make it possible to understand the pathological effects of expressing human Aβ42 peptides on the ultrastructure of the brain. Immunoelectron microscopy enables correlation of the subcellular distribution of Aβ42 peptides with cellular damage at the ultrastructural level, which provides insights into the mechanisms of Aβ42 toxicity in vivo. Specifically localizing Aβ42 peptides using the electron microscope with post-embedding immunogold labeling is best performed on samples treated with osmium tetroxide so that cell membranes can be visualized. The protocol presented above uses relatively simple methods of sample preparation and allows strong and specific labeling on thin sections for Aβ42 in both neuronal and glial cells.

Figure 6 shows the distribution of human Aβ42 expressed in the transgenic fly brain [13, 14]. Immunoelectron microscopy-detected Aβ42 is observed in the endoplasmic reticulum, Golgi, and lysosomes (Fig. 6a, b). The Aβ42 accumulation is also occasionally detected in glial cells, suggesting that Aβ42 peptides are secreted from neurons and then taken up by glial cells (Fig. 6c). Gold particles are absent in the control section, which is a fly brain that does not express Aβ42 (Fig. 6d).

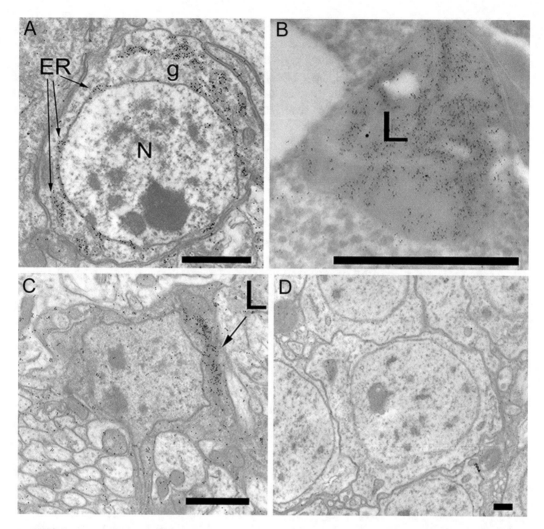

Fig. 6 Examples of immuno-EM analyses of Aβ in the fly brain. (**a**) ImmunoEM detection of Aβ in the endoplasmic reticulum (ER) and Golgi, as well as (**b**) a lysosome (L) in the neurons in Kenyon cell region of Aβ fly brains. (**c**) Immunogold labeling for Aβ over a lysosome in a glial cell (*arrowhead*). (**d**) Gold particles are absent in section labeled for Aβ in the brain of a control fly that does not express Aβ. Scale bars all equal 1 μm. *N* nucleus

4   Conclusion

EM analyses are powerful methods that are used to identify cellular phenotypes and localize proteins to subcellular organelles at the ultrastructural level, which can be critical for validating animal models of diseases. Indeed, application of EM to the brains of flies modeling AD revealed that the cellular phenotypes reported in patients with AD, such as necrosis, neurite degeneration, and abnormalities in mitochondria and lysosomes, are recapitulated. Furthermore, EM analyses in fly models of AD suggest that neurodegeneration can occur in the absence of fibrillar forms of Aβ or tau [13, 16]. Thus, the use of EM analyses in animal models of disease can provide not only a description of the morphological changes but also insight into the mechanisms underlying the disease, which may then be directly tested in these animal models.

Acknowledgement

This work was supported by grants from the National Institute of Health [R01AG032279] and [U01AG046170] to K.A. and K.M.I., the Alzheimer's Association NIRG-10-173189 to K.A. and NIRG-08-91985 to K.M.I., the start-up funds from Tokyo Metropolitan University (to K.A.), Takeda Science Foundation, Japan (to K.M.I.), and Research Grant for Longevity Science 25-27, Japan (to K.M.I.).

References

1. Reiter LT, Potocki L, Chien S, Gribskov M, Bier E (2001) A systematic analysis of human disease-associated gene sequences in Drosophila melanogaster. Genome Res 11 (6):1114–1125

2. Fortini ME, Skupski MP, Boguski MS, Hariharan IK (2000) A survey of human disease gene counterparts in the Drosophila genome. J Cell Biol 150(2):F23–F30

3. Ambegaokar SS, Roy B, Jackson GR (2010) Neurodegenerative models in Drosophila: polyglutamine disorders, Parkinson disease, and amyotrophic lateral sclerosis. Neurobiol Dis 40(1):29–39

4. Newman T, Sinadinos C, Johnston A, Sealey M, Mudher A (2011) Using Drosophila models of neurodegenerative diseases for drug discovery. Exp Opin Drug Discov 6(2):129–140

5. Iijima-Ando K, Iijima K (2010) Transgenic Drosophila models of Alzheimer's disease and tauopathies. Brain Struct Funct 214 (2–3):245–262

6. Cummings JL (2003) The neuropsychiatry of Alzheimer's disease and other dementias. Martin Dunitz, London

7. Selkoe DJ (2002) Alzheimer's disease is a synaptic failure. Science 298(5594):789–791

8. Selkoe DJ (2001) Alzheimer's disease: genes, proteins, and therapy. Physiol Rev 81 (2):741–766

9. Glenner GG, Wong CW (1984) Alzheimer's disease and Down's syndrome: sharing of a unique cerebrovascular amyloid fibril protein. Biochem Biophys Res Commun 122 (3):1131–1135

10. Masters CL, Simms G, Weinman NA, Multhaup G, McDonald BL, Beyreuther K (1985) Amyloid plaque core protein in Alzheimer disease and Down syndrome. Proc Natl Acad Sci U S A 82(12):4245–4249

11. Lee VM, Balin BJ, Otvos L Jr, Trojanowski JQ (1991) A68: a major subunit of paired helical filaments and derivatized forms of normal Tau. Science 251(4994):675–678

12. Hardy J, Selkoe DJ (2002) The amyloid hypothesis of Alzheimer's disease: progress and problems on the road to therapeutics. Science 297(5580):353–356

13. Iijima K, Liu HP, Chiang AS, Hearn SA, Konsolaki M, Zhong Y (2004) Dissecting the pathological effects of human Abeta40 and Abeta42 in Drosophila: a potential model for Alzheimer's disease. Proc Natl Acad Sci U S A 101(17):6623–6628

14. Iijima K, Chiang HC, Hearn SA, Hakker I, Gatt A, Shenton C, Granger L, Leung A, Iijima-Ando K, Zhong Y (2008) Abeta42 mutants with different aggregation profiles induce distinct pathologies in Drosophila. PLoS One 3(2):e1703

15. Finelli A, Kelkar A, Song HJ, Yang H, Konso-laki M (2004) A model for studying Alzheimer's Abeta42-induced toxicity in Drosophila melanogaster. Mol Cell Neurosci 26 (3):365–375

16. Wittmann CW, Wszolek MF, Shulman JM, Salvaterra PM, Lewis J, Hutton M, Feany MB (2001) Tauopathy in Drosophila: neurodegeneration without neurofibrillary tangles. Science 293(5530):711–714

17. Jackson GR, Wiedau-Pazos M, Sang T-K, Wagle N, Brown CA, Massachi S, Geschwind DH (2002) Human wild-type Tau interacts with wingless pathway components and produces neurofibrillary pathology in Drosophila. Neuron 34(4):509–519

18. Iijima-Ando K, Sekiya M, Maruko-Otake A, Ohtake Y, Suzuki E, Lu B, Iijima KM (2012) Loss of axonal mitochondria promotes Tau-mediated neurodegeneration and Alzheimer's disease-related Tau phosphorylation via PAR-1. PLoS Genet 8(8):e1002918

19. Chee FC, Mudher A, Cuttle MF, Newman TA, MacKay D, Lovestone S, Shepherd D (2005) Over-expression of tau results in defective synaptic transmission in Drosophila neuromuscular junctions. Neurobiol Dis 20(3): 918–928

20. Nishimura I, Yang Y, Lu B (2004) PAR-1 kinase plays an initiator role in a temporally ordered phosphorylation process that confers tau toxicity in Drosophila. Cell 116 (5):671–682

21. Milyaev N, Osumi-Sutherland D, Reeve S, Burton N, Baldock RA, Armstrong JD (2012) The Virtual Fly Brain browser and query interface. Bioinformatics 28(3):411–415

22. Brand AH, Perrimon N (1993) Targeted gene expression as a means of altering cell fates and generating dominant phenotypes. Development 118(2):401–415

23. Gibson GE, Nielsen P, Sherman KA, Blass JP (1987) Diminished mitogen-induced calcium uptake by lymphocytes from Alzheimer patients. Biol Psychiatry 22(9):1079–1086

24. Zempel H, Mandelkow E (2014) Lost after translation: missorting of Tau protein and consequences for Alzheimer disease. Trends Neurosci 37(12):721–732

25. Benilova I, Karran E, De Strooper B (2012) The toxic Abeta oligomer and Alzheimer's disease: an emperor in need of clothes. Nat Neurosci 15(3):349–357

26. Palade GE (1952) A study of fixation for electron microscopy. J Exp Med 95(3): 285–298

27. Hernandez-Moran H (1957) Electron microscopy of nervous tissue. Metabolism of the nervous system. Pergamon, London

28. Bendayan M, Zollinger M (1983) Ultrastructural localization of antigenic sites on osmium-fixed tissues applying the protein A-gold technique. J Histochem Cytochem 31(1):101–109

29. Hearn SA, Silver MM, Sholdice JA (1985) Immunoelectron microscopic labeling of immunoglobulin in plasma cells after osmium fixation and epoxy embedding. J Histochem Cytochem 33(12):1212–1218

30. Yamashita S, Okada Y (2014) Heat-induced antigen retrieval in conventionally processed epon-embedded specimens: procedures and mechanisms. J Histochem Cytochem 62 (8):584–597

31. Mathiisen TM, Nagelhus EA, Jouleh B (2006) Postembedding immunogold cytochemistry of membrane molecules and amino acid transmitters in the central nervous system. Neuroanatomical tract-tracing 3. Springer Press eBook, New York, pp 72–108

32. Merighi A, Polak JM, Fumagalli G, Theodosis DT (1989) Ultrastructural localization of neuropeptides and GABA in rat dorsal horn: a comparison of different immunogold labeling techniques. J Histochem Cytochem 37 (4):529–540

33. Zago W, Schroeter S, Guido T, Khan K, Seubert P, Yednock T, Schenk D, Gregg KM, Games D, Bard F, Kinney GG (2013) Vascular alterations in PDAPP mice after anti-Abeta immunotherapy: implications for amyloid-related imaging abnormalities. Alzheimers Dement 9(5 Suppl):S105–S115

34. Amiry-Moghaddam M, Ottersen OP (2013) Immunogold cytochemistry in neuroscience. Nat Neurosci 16(7):798–804

35. Van Dorpe J, Smeijers L, Dewachter I, Nuyens D, Spittaels K, Van Den Haute C, Mercken M, Moechars D, Laenen I, Kuiperi C, Bruynseels K, Tesseur I, Loos R, Vanderstichele H, Checler F, Sciot R, Van Leuven F (2000) Prominent cerebral amyloid angiopathy in transgenic mice overexpressing the London mutant of human APP in neurons. Am J Pathol 157 (4):1283–1298

36. Yu WH, Kumar A, Peterhoff C, Shapiro KL, Uchiyama Y, Lamb BT, Cuervo AM, Nixon RA (2004) Autophagic vacuoles are enriched in amyloid precursor protein-secretase activities: implications for beta-amyloid peptide over-production and localization in Alzheimer's disease. Int J Biochem Cell Biol 36(12): 2531–2540

Neuromethods (2016) 115: 125–138
DOI 10.1007/7657_2015_80
© Springer Science+Business Media New York 2015
Published online: 18 July 2015

A Multifaceted Approach with Light and Electron Microscopy to Study Abnormal Circuit Maturation in Rodent Models

Kimberly L. Simpson, Yi Pang, and Rick C.S. Lin

Abstract

In order to better understand circuit function and dysfunction during early brain formation, multiple approaches at the light and electron microscopic levels have been utilized to greatly enhance our knowledge regarding the manner in which neurons and members of the glial cell family develop and respond to insult. Here, we document the use of immunofluorescent staining methods in addition to ultrastructural analysis to reveal selective alteration of chemoarchitectural substrates in both neurons and glial cells after early exposure to an antidepressant, a neurotoxin, as well as genetic manipulation. Our novel findings have vast implications for numerous research fields including studies related to neurological disease, mental disorders, as well as pervasive developmental disorders, such as autism.

Keywords: Serotonin, Norepinephrine, Dopamine, Development, Dendrites, Axons, Oligodendrocytes, Microglia, Injury

1 Introduction

The development of a normal functioning brain not only requires a structure-specific framework of local and long-range connections but also depends upon specific neurochemicals and the precise timing of their release.

It is now well recognized that as the brain reconciles new internal and external environments or challenging conditions, it has the capacity to undergo remarkable architectural changes and learns to adjust to new environments or conditions.

With the generation of rather selective targeting techniques and enhanced methodologies for the visualization of cellular and subcellular profiles (via improvements in the applicability of receptor-specific antibodies and refinement of microscopic technology for the capture of high-resolution digitized fluorescent and electron images), neuroscientists now have the ability to probe early events in brain formation and evaluate mechanisms that perturb the normal trajectory of pathway establishment. Combined utilization of these improved methodologies has, indeed, permitted us the opportunity to assess aberrant brain development in response to genetic manipulations and early exposure to environmental factors.

At present, most discussions relating abnormal anatomical constructs to stunted brain development have cited alterations in the configuration and operation of excitatory glutamatergic and inhibitory GABAergic circuits. In contrast, relatively little is known about the contributions of other neurotransmitter families, particularly substances considered to act as neuromodulators. Interestingly, biogenic amine neurotransmitters including dopamine, serotonin (5-hydroxytryptamine; 5-HT), and norepinephrine (NE) have been shown to exert an important role in early development through the regulation of neurogenesis, migration, differentiation, and plasticity [1–4]. In fact, manipulation of 5-HT during early development has been demonstrated to interfere with whisker barrel formation in the rodent primary somatosensory cortex, produce over-reaction to sensory stimulation, as well as lead to aggressive and/or anxiety-related behaviors [5–11]. Furthermore, 5-HT can be transiently expressed in numerous nonserotonergic neurons such as primary sensory thalamic neurons, dopaminergic substantia nigra (SN) neurons, and even noradrenergic locus coeruleus (LC) neurons during the first few weeks of life in the rodent brain [12, 13], thereby suggesting a potential diversity of dysfunction after initial, early disruption of the serotonin system. Clinical studies which support this view document serotonin dysregulation to be linked to a variety of mental disorders such as depression, anxiety, aggression, obsessive compulsive disorder, and even autism [14, 15]. These findings correlate well with animal investigations which demonstrate early exposure to SSRI antidepressants can lead to extensive cortical mis-wiring and abnormal behavior in the rodent, reminiscent of pathophysiology noted in autism [16]. Likewise, circumstances which promote perinatal infection/inflammation increase the risk for developing disorders related to dopaminergic neurons, such as Parkinson disease, as well as schizophrenia, autism, and cerebral palsy [17–20].

In brief, this new knowledge indicates that neuromodulators may play a critical role in early development. In the following section, we will describe outcomes in terms of neuronal and glial structural changes. Several rodent animal models are considered, including a model which leverages the use of antidepressants to selectively inhibit the uptake of serotonin (SSRIs) early in brain formation. In addition, models which employ early administration of neurotoxins such as lipopolysaccharide (LPS) as well as manipulation of monoamine oxidase A through gene knockout are also discussed.

2 Materials and Methods

2.1 Perinatal Exposure to Antidepressants

Male and female offspring of timed-pregnant Long Evans rats were used in these experiments. On postnatal days 6–7 (PN6-7), pups were tattooed and sexed for identification. Beginning on PN8, rat pups were injected subcutaneously with citalopram (CTM) (Tocris, Ellisville, MO) at a dose ranging from 2.5 to 10 mg/kg twice daily throughout the PN8-21 time frame. Investigations to examine pathology were conducted when animals reached adulthood (>PN90). For immunohistochemical studies, animals were deeply anesthetized with sodium pentobarbital (60–80 mg/kg; IP) and perfused through the heart with saline followed by 3.5 % paraformaldehyde in 0.1 M phosphate-buffered saline (PBS) at pH 7.4. Free floating sections were collected at 40–80 μm thickness with an AO freezing microtome. Standard immunofluorescent methods were conducted as described previously [21, 22]. For example, tryptophan hydroxylase (TPH) affinity purified antiserum against sheep (1:1,000–3,000; Millipore Co.) was used to identify serotonergic neurons in the raphe nuclear complex, and linked to either streptavidin-conjugated Cy2 or Cy3. The immunostained tissues were then examined for their distribution pattern and intensity utilizing a Nikon E800 microscope equipped with appropriate filters and Photometric Coolsnap ES CCD camera (Roper Scientific Co.), as well as Metamorph imaging software (Universal imaging systems) for quantification. This newly developed quantification system allows an investigator to not only measure the intensity of staining but also detect the spatial arrangement of cellular profiles [23].

Brain samples for electron microscopic (EM) investigation were also prepared with a standard ABC/DAB immunohistochemical ultrastructural procedure. Typically, animals were perfused with saline followed by 3.5 % paraformaldehyde plus 0.5 % glutaraldehyde in 0.1 M PBS. Brains were post-fixed in the same fixative overnight at 4 °C. Brains were cut either in the coronal (for raphe neurons and Nodes of Ranvier) or sagittal (for corpus callosum studies) plane. A Lancer vibratome was used to slice sections to 50–100 μm. A small region (~1 × 1 mm) of interest was dissected with a #10 blade. Typically, 2–3 blocks were dissected from each case and processed with standard EM osmication with en bloc staining procedures. They were flat embedded in epon, attached to beam capsules, trimmed and cut into ultrathin sections. Tissues were collected onto grids coated with formvar, and further stained with lead citrate and uranyl acetate. Materials were examined, randomly selected, and photographed with a Leo Biological transmission electron microscope equipped with a digitized imaging system. A Jeol JEM 1400 Plus EM scope was also used for these studies.

2.2 Monoamine Oxidase A Knockout (MAO A KO) Experiments

MAO A KO mice were generated as described previously [24, 25]. Briefly, MAO A KO mice were engineered from the 129S6 (WT) strain. This breed harbors a spontaneous point nonsense mutation in exon 8, position 863, of the *Maoa* gene. These MAO A KO mice lack MAO A enzyme activity and have a high level of 5-HT throughout the brain [24].

With regard to immunofluorescent light microscopic and ultrastructural analysis, similar procedures were conducted as described in section 2.1.

2.3 Perinatal Exposure to Neurotoxin

To investigate neurotoxin effects during early development, timed-pregnant Sprague–Dawley rats were used. Intraperitoneal injection of LPS (from *Escherichia coli*, serotype 055:B5 at a dose of 0–3.0 μg/g or 0–9,000 EU/g body weight) was administered at PN5. Rat pups were then weaned at PN21 and were housed in 3–4 per cage thereafter. To further assess toxic effects, rotenone (a pesticide) was administered via subcutaneous osmotic minipump infusion for 14 days (from PN70 to PN84; 1.25 mg/kg/day) as described previously [26]. These animals were then further studied for degenerative processes at >PN98.

For immunohistochemical and ultrastructural investigation, methods were conducted as described in the previous section. With regard to the usage of antibodies, tyrosine hydroxylase primary antiserum (TH; dilution 1:1,000; Millipore Co.) was used to identify dopaminergic neurons in the substantia nigra (SN), and OX42 (CD11b; dilution 1:100; Serotec Co.) or ionized calcium-binding adapter molecule 1 (Iba1; dilution 1:200; Wako Co.) antiserum was used to identify microglia, and Contactin-associated protein (Caspr: dilution 1:200; Chemicon Inc.) antiserum was used to identify nodal subdomain of the myelination. Furthermore, dendritic morphological changes as well as microglia pathology were studied through the use of electron microscopy.

2.4 In Vitro Cell Culture Experiments

To further verify and understand mechanisms of degeneration in specific populations of neurons and glia, in vitro cell culture, including the co-culture method, was conducted and cellular profiles were examined at both the light and electron microscopic levels. Oligodendrocyte progenitor cells were isolated from neonatal rat optic nerves as described previously [27]. Co-culture material derived from embryonic rat spinal cord and cerebral cortex was harvested to further study neuron-glia interaction. Rat E16 spinal cord was dissected from six embryos and collected in a petri dish containing 1 ml of 1× HBSS (without Ca^{++} and Mg^{++}). Additional procedures have been described recently [28].

Ultrastructural organization in these neuron-glia co-cultures was characterized by a new method which serves to evaluate the extent and patterns of myelination [28]. Briefly, the dissociated cells in the plating medium were seeded into Matrigel Matrix Cell Culture inserts (BD Biosciences, Bedford, MA) at the same density as that on the cover slips. At DIV 40, cells were fixed with 0.5 % glutaraldehyde for 30 min at room temperature, washed and stored in PBS at 4 °C, and then processed utilizing standard EM osmication with en bloc staining of 2 % uranyl acetate for 5 min. The tissues were embedded in Durcupan and ultrathin sections were cut and examined with a Leo Biological transmission electron microscope.

3 Results

3.1 Alterations in Dendritic Morphology Following Early Exposure to Antidepressant Drugs, Inflammatory Agents, or Genetic Manipulations

A typical neuron is composed of a cell body or soma, dendrites, an axon, and associated terminal synapses. In responding to an external insult such as ischemia, degradation of a dendritic structural protein has been reported. Matesic and Lin [29] observed that hippocampal dendrites exhibit beading very early after five minute carotid artery occlusion. Namely, a change in the integrity of microtubule-associated protein 2 was detected as early as 24 h post-insult and occurred in advance of neuronal death. With sublethal insult, however, such structural alterations of dendrites were found to be reversible [30]. To explore the possibility of similar permutations of dendritic structure within monoaminergic systems following early exposure to antidepressant drugs, inflammation, or genetic manipulation, three different experimental models were analyzed. The first model utilized intracerebroventricular or intraperitoneal injection of LPS at PN5. Irrespective of the route of delivery, by the time subjects reached adulthood, substantia nigra cells labeled immunohistochemically for TH demonstrated shortened or beaded dendrites in comparison to the smooth dendritic profiles exhibited by controls. At the ultrastructural level, SN dendrites of LPS-treated subjects demonstrated areas that were vacant of content, and contained numerous vacuoles. In many instances, internalized myelinated axons were observed (Fig. 1a–c). Such alterations appear much more severe following a second sublethal insult with rotenone (Fig. 1d). Secondly, in models examining the expression of TPH after early exposure to the SSRI antidepressant, citalopram, midline raphe neurons tended to express very weak or no immunoreactivity. Interestingly, some of the TPH immunoreactive neurons of dorsal raphe (DR) neurons from CTM-exposed animals exhibited a beaded appearance that contrasted with the rather smooth dendrites of controls (Fig. 1e). Such morphological changes were reinforced by data at the EM level. Again, inclusions and vacuoles were noted in TPH immunoreactive DR dendrites after perinatal CTM exposure (Fig. 1f from normal and G from CTM treated). Lastly, to explore consequences following genetic manipulation, immunofluorescent data from MAO A KO mice revealed an altered expression of TH in the dendrites of locus coeruleus (LC) noradrenergic neurons. Processes exhibited a beaded appearance in a time-dependent fashion such that dendritic

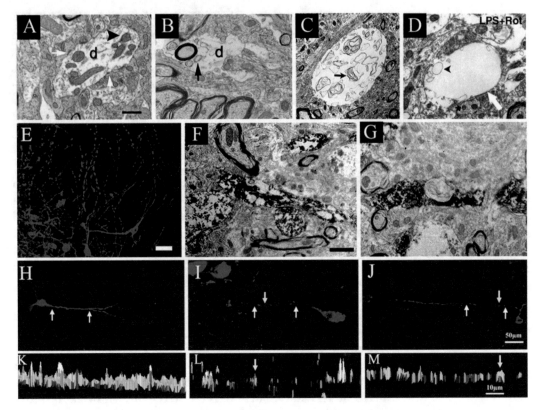

Fig. 1 Dendritic pathology following perinatal drug exposure. Representative example of the normal appearance of a dendrite following saline treatment (P5 treated and examined at adult) (**a**: *arrow head* indicates normal mitochondria). In contrast, dendrites of LPS treated at P5 animals were found to be largely vacant of content, but contain numerous vacuoles and even internalized myelinated axons (examined at adult) (**b** and **c**: *arrows*). Further treatment with rotenone appears to worsen dendritic pathology (**d**: *arrow head* points to a rather empty vacuole). Dendritic abnormalities related to beading of TPH immunostained dendrites were observed in the dorsal raphe nuclear complex (**e**), with alterations also appearing at the ultrastructural level (**f**: normal); and (**g**) following perinatal exposure to CTM. Dendritic labeling in the LC is shown in **h–j** (*white arrows* point to TH immunoreactive processes). LC neurons exhibited rather smooth dendrites in normal controls (normal adult: **h**), but the dendrites of MAO KO mice demonstrated a beaded appearance at PN 30 (**i**). Such discontinuity in the integrity of dendritic branches appeared to improve in part by PN 150 (**j**). The staining intensity and spatial distribution of the TH immunoreactive profiles shown in **h–j** are shown high power in **k–m** (*white* indicates TH labeled areas with high signal intensity and *blue* denotes TH labeled areas with low signal intensity). *Yellow arrows* (**i–j** and **l–m**, respectively) point to individual beads along the extent of the dendrite and relative levels of intensity associated with that beaded profile. Scale bar: (**a**) 1 μm, also applies to (**b–d**); (**e**) 25 um; (**f**) 2 μm also applies to (**g**)

branches demonstrated discontinuous arborization at (P30), but returned to normal by (P150) (Fig. 1h–m). In brief, these new lines of evidence provide further support to demonstrate a link between altered dendritic morphology and a general process that has the potential to disrupt neuronal outflow and lead to degeneration. At present, the precise cellular mechanism responsible for these changes remains to be elucidated, as does a means for limiting the occurrence of this phenomenon.

3.2 Morphological Changes in the Myelination of Corpus Callosum Axons Following Early Drug Exposure and Genetic Manipulation

As was stated earlier, serotonin is known to play a key role in axon guidance and cortical barrel formation during early development. Given clinical reports of decreased corpus callosum volume in autistic patients [31], we speculated that early manipulation of serotonin would lead to aberrant axon myelination and further evidence of misguidance during the establishment of neuronal pathways. To illustrate myelin malformation at the ultrastructural level, we studied rat pups that had received early CTM treatment as well as rat pups with altered expression of MAO A. Under the first condition, where subjects received early exposure to the SSRI either pre- or postnatally, data revealed that a portion (2–6 %) of callosal axon myelin sheaths was often rather loosely packed (21–60 abnormal axons from a sample of 1,000 vs. 5–9 abnormal axons from the same sample size from the controls) [16]. On occasion, hypermyelination was also noted as compared to the nicely organized and tightly packed lamellae of controls (Fig. 2a–c). When callosal axon myelin formation was investigated in MAO A KO mice, abnormal patterns of sheathing were quite often noted (Fig. 2d–f). In some instances, multiple axons with abnormally organized lamellae were found wrapped together by another giant myelin sheath that was also anatomically distorted. These new lines of evidence not only support the view that early environment or genetic manipulation can re-program pathway formation during development but also

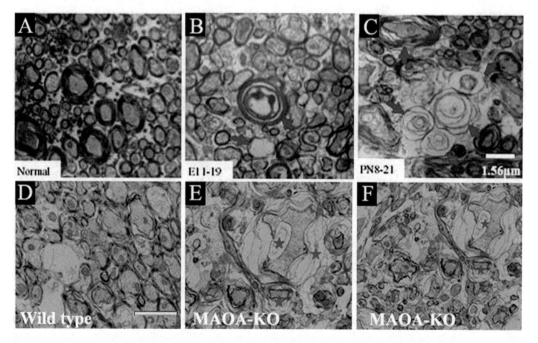

Fig. 2 Callosal axons at the ultrastructural level. Early CTM exposure leads to the abnormal appearance of callosal axons (**b** and **c**) compared to the normal myelinated axons of saline-treated animals (**a**). The *red arrows* point to several representative examples of loosely packed myelin sheaths. Similar findings were also observed in MAO A KO mice (**e** and **f**: the *asterisk* marks the abnormal appearance of callosal axons). Scale bar: (**d**) 2 μm, also applied to (**e** and **f**)

further suggest that alterations in serotonin availability can interfere with the assembly of neuronal machinery responsible for reliable signal transmission.

3.3 Altered Node of Ranvier

Since callosal axons did not demonstrate organized patterns of myelination in experimental models of altered serotonin homeostasis, steps were taken to explore subdomains of the Node of Ranvier. Coronal sections of the corpus callosum were collected to maximize the view of the Node of Ranvier, and architectural changes were investigated with the use of both immunofluorescent and electron microscopic techniques. The Caspr antibody was used to immunohistochemically identify the paranodal area and evaluate paranodal integrity. In control experiments, the typical space separating two Caspr immunostained profiles was quite narrow (range of ~1–2 μm). In contrast the perinatal CTM-exposed animals exhibited a rather large gap (3–10 μm) between Caspr immunostained structures (Fig. 3a, b), suggesting an additional dimension of malformation relating to axon myelination. To further substantiate such structural changes, investigation at the electron microscopic level revealed numerous incidences of an extensive naked zone (~5–10 μm) without any of the obvious end-feet subdomains that were noted in normal controls (Fig. 3c–f).

3.4 Oligodendrocyte Pathology

Since the myelin sheath is formed by oligodendrocytes (OL), and abnormal myelin sheath formations were detected in animals following manipulation of serotonin levels during early development,

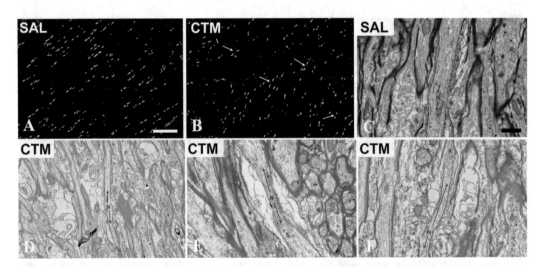

Fig. 3 Node of Ranvier pathology following perinatal exposure to CTM. Paranodal Caspr immunostained profiles reveal a rather regular and close spacing between two stained paranodes in saline-treated controls (**a**). In contrast, CTM-treated subjects demonstrated larger distances and inconsistent spacing between Caspr stained profiles (**b**: *white arrows* point to representative examples). Samples examined at the ultrastructural level show normal nodal spacing following saline treatment (**c**) and disfigured nodal domains following CTM administration (**d–f**); the *rows* indicate the length of the node (N) region. *P* paranodal region. Scale bar: (**a**) 10 μm applies also to (**b**); (**c**) 2 μm applies also to (**d, e**, and **f**)

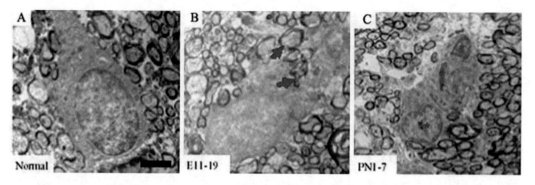

Fig. 4 Ultrastructural analysis of oligodendrocyte pathology. Representative examples of normal saline-treated controls are shown in (**a**) while (**b** and **c**) illustrate oligodendrocyte somata from CTM perinatally treated animals. Notice inclusion bodies such as myelinated axons (**b**) and evidence of multinucleation (**c**: indicated by *asterisks*). Scale bar: (**a**) 5 μm, also applies to (**b**) and (**c**)

an obvious next step was to use EM to examine the disposition of OLs in experimental subjects. After perinatal exposure to CTM, OL cell bodies in the corpus callosum were found to contain numerous vacuoles and lysosome inclusions. The existence of such inclusions was not noted in controls (Fig. 4a–c). Furthermore, myelinated axonal profiles and multinucleated OLs were also noted in the soma from our samples. Interestingly, the pathological appearance of OLs in CTM-treated material was similarly observed in MAO A KO mice. This information suggests that OLs may sustain damage during developmental periods when serotonin is abnormally regulated.

3.5 Microglial Reaction to Early Insult

It has been well documented that microglia play a critical role in responding to insult in the central nervous system. In particular, both neurological and psychiatric disorders in children and adults have been linked to prenatal infection/inflammation [32–34]. Animal studies which have sought to characterize the putative impact of microglial activation in response to perinatal LPS exposure have shown degeneration of dopaminergic neurons in the substantia nigra (SN) and activated microglia in the same region [17, 35, 36]. Interestingly, sequential exposure to a sublethal dose of another neurotoxin can promote further damage in this same population of neurons, even when the secondary treatment is administered in adult subjects [35, 36], suggesting a unique vulnerability present in dopaminergic circuitry. Reactive microglia immunostained with Iba1, under these conditions, exhibited lysosome inclusions, enlarged mitochondria, and extended cytoplasmic processes (Fig. 5a–c). Since early SSRI exposure appears to lead to abnormal circuit wiring and malformation of the myelin sheath, future investigations will address whether microglia are also activated in response to serotonin dysregulation. Our immunofluorescent data suggest that perinatal exposure to CTM induces extension of microglia cytoplasmic

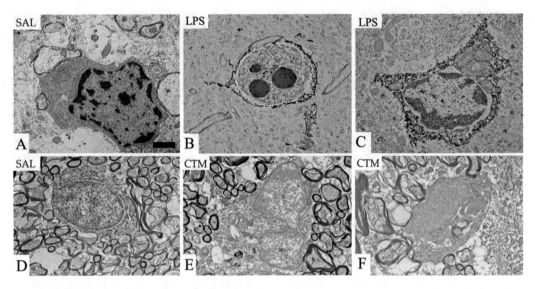

Fig. 5 Microglia pathology following early life exposure to an inflammatory drug or SSRI. Iba immunostained microglia in the hippocampus from saline-treated controls (**a**) and perinatal LPS-exposed subjects (**b** and **c**). Notice the lysosomal inclusions and extensive numbers of mitochondria in the drug-treated group. There were many instances in CTM-treated subjects in the corpus callosum where microglia exhibited abnormalities such as multinucleation (**e**: designated by *red asterisks*) and inclusion bodies/vacuoles (**f**: marked by *green asterisks*). Scale bar: (**a**) 2 μm, applies also to (**b–f**)

processes and enhances soma size. Some of these observations at the light microscopic level have been further identified utilizing electron microscopy (Fig. 5d–f).

3.6 Myelination and Synaptic Organization Utilizing the Co-culture Method

The conventional pure OL culture has been well established for decades. However, the co-culture model which utilizes both OL and neurons to investigate myelin formation and synaptic profiles has not been widely available. Dorsal root ganglion material was most often used in the absence of constituents liberated from the CNS [37, 38]. To develop an approach which incorporated material originating from the CNS, Yang et al. [39] and Chen et al. [40] introduced the brain slice and explant culture, respectively. Limitations which were associated with such CNS-based models, however, were well recognized [41]. With the advent of a newly developed co-culture method [28], we have been able to determine the process of OL maturation and myelination by immunohistochemistry using OL, myelin, or nodal domain markers (Fig. 6a–d). In particular, ultrastructural evidence of myelin formation as well as synaptic specializations was observed (Fig. 6e–h). Furthermore, pre- and postsynaptic features as well as dense-cored vesicles were noted in the newly developed co-culture system (Fig. 6f–h), thereby suggesting the feasibility of using EM analysis to study co-culture systems in response to a variety of experimental manipulations.

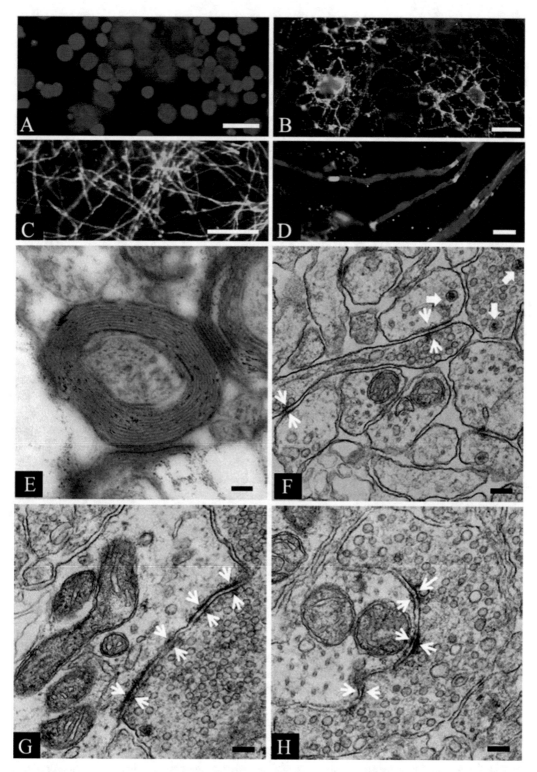

Fig. 6 Myelination and synaptic organization in mixed neuron-glia co-culture derived from E16 spinal cord. Representative micrographs show neurons (**a**: *red*, NeuN staining; nuclei were counterstained *blue*, DAPI) and

4   Notes and Discussion

4.1   Technical Considerations

The major focus of the present set studies was to emphasize the importance of applying multiple approaches to not only advance understanding of degenerative processes but also appreciate how seemingly subtle changes in the architecture of monoamine systems (traditionally viewed as neuromodulatory systems) can alter the trajectory of brain development. The current collection of studies has also been presented to reinforce the notion that the glial cell family, especially oligodendrocytes and microglia, may also play a prominent role in impacting dysfunction associated with altered levels of serotonin, dopamine, and potentially, norepinephrine. The combined use of immunohistochemistry at the light microscopic level with translation at the ultrastructural level offers researchers an in-depth view of synaptic relationships and an opportunity to categorize domains of anatomic deficiency.

4.2   Functional Implications

The importance of investigating dendritic integrity is underscored by a need to better understand frameworks for functional circuit connections. Given that the majority of synaptic interactions are built upon axonal contacts with dendrites and dendritic spines, it stands to reason that a breakdown in the dendritic construct could have a pronounced detrimental effect on neuronal activity and survival. As was stated earlier, dendritic beading of microtubule-associated protein 2 was observed in hippocampal neurons following ischemic insult [29]. This damage was shown to predict degeneration with longer durations of carotid artery occlusion. The cellular mechanism mediating this phenomenon was found to involve reduced voltage-dependent Ca^{++} signaling [42]. A focus for the future will be to establish whether dendritic beading noted in monoamine neurons following early exposure to antidepressants and inflammatory agents, as well as genetic manipulation, is also mediated through similar cellular mechanisms (i.e., calcium deprivation prior to cell death).

Fig. 6 (continued) immature OLs (**b**: *green*, O4 immunostaining; neuronal processes were immunostained *red* with Tuj1) at early stages of the culture (DIV10). (**c–d**) Myelinated axons at later stages (DIV40) were revealed via double-immunolabeling with myelin basic protein (MBP) and phosphorylated neurofilament H (**c**: *green* and *red* fluorescence, respectively), which was mostly detected at late stage (DIV40 as shown in **c**). Double immunolabeling of MBP and the protein Caspr in the nodal domain (**d**: red and green fluorescence, respectively) revealed that the cytoarchitecture of myelinated axons in cell culture is similar to that found in vivo. (**e–h**) Ultrastructural demonstration of myelin and synaptic organization in mixed neuron-glia culture. A representative EM image shows that compact myelin sheaths form around axons under normal culture condition (**e**). A variety of synaptic specializations, such as synaptic membranes with electron dense (indicated by *opposing arrows*) and dense-cored vesicles (denoted by *thick white arrows*), were noted in the culture (**f, g, h**). Scale bars: (**a–c**) 25 µm; (**d**) 10 µm; (**e–f**) 100 nm

4.3 Future Directions

There are several directions that can be followed from the present data. For instance, based on the newly developed co-culture model, there is an option to evaluate whether hypomyelination and/or demyelination can be induced following exposure to proinflammatory cytokines, such as IL-1β or TNFα. Also demyelinating agents such as lysophosphatidylcholine (LPC) as well as a combination of normal guinea pig serum with anti-MOG (myelin oligodendrocyte glycoprotein) antibody [28] can also be explored. At present, such investigations are moving forward at the light microscopic level using fluorescence immunohistochemistry. Future advances along this front are anticipated in upcoming EM studies.

References

1. Azmitia EC (2001) Modern view on ancient chemical: serotonin effects on proliferation, maturation, and apoptosis. Brain Res Bull 56:414–424

2. Gaspar P, Cases O, Maroteaux L (2003) The developmental role of serotonin: news from mouse molecular genetics. Nat Rev Neurosci 4:1002–1012

3. Marder E (2012) Neuromodulation of neuronal circuits: back to the future. Neuron 76:1–11

4. Thompson BL, Stanwood GD (2009) Pleiotropic effects of neurotransmission during development: modulators of modularity. J Autism Dev Disord 39:260–268

5. Cases O et al (1995) Aggressive behavior and altered amounts of serotonin and norepinephrine in mice lacking MAOA. Science 268:1763–1766

6. Cases O et al (1996) Lack of barrels in the somatosensory cortex of monoamine oxidase A-deficient mice: role of a serotonin excess during the critical period. Neuron 16:297–307

7. Persico AM et al (2000) Serotonin depletion and barrel cortex development: impact of growth impairment vs. serotonin effects on thalamocortical endings. Cereb Cortex 10:181–191

8. Persico AM et al (2001) Barrel pattern formation requires serotonin uptake by thalamocortical afferents, and not vesicular monoamine release. J Neurosci 21:6862–6873

9. Holmes A, Yang RJ, Lesch KP, Crawley JN, Murphy DL (2003) Mice lacking the serotonin transporter exhibit 5-HT1A receptor-mediated abnormalities in tests for anxiety-like behavior. Neuropsychopharmacology 28:2007–2088

10. Xu Y, Sari Y, Zhou FC (2004) Selective serotonin reuptake inhibitor disrupts organization of thalamocortical somatosensory barrels during development. Brain Res Dev Brain Res 150:151–161

11. Jennings KA et al (2006) Increased expression of the 5-HT transporter confers a low-anxiety phenotype linked to decreased 5-HT transmission. J Neurosci 26:8955–8964

12. Lebrand C et al (1996) Transient uptake and storage of serotonin in developing thalamic neurons. Neuron 17:823–835

13. Narbour-Neme N, Pavone LM, Avallone L, Zhuang XX, Gaspar P (2008) Serotonin transporter transgenic (SERT cre) mouse line reveals developmental targets of serotonin specific reuptake inhibitors (SSRIs). Neuropharmacology 55:994–1005

14. Lucki I (1998) The spectrum of behaviors influenced by serotonin. Biol Psychiatry 44:151–162

15. Anderson GM (2004) Peripheral and central neurochemical effects of the selective serotonin reuptake inhibitors (SSRIs) in humans and nonhuman primates: assessing bioeffect and mechanisms of action. Int J Dev Neurosci 22:397–404

16. Simpson KL, Weaver KJ, de Villers-Sidani E, Lu JYF, Cai ZW, Pang Y, Rodriquez-Porcel F, Paul IA, Merzenich M, Lin RCS (2011) Perinatal antidepressant exposure alters cortical network function in rodents. Proc Natl Acad Sci 108:18465–18470

17. Ling Z, Gayle DA, Ma SY, Lipton JW, Tong CW, Hong JC, Carvey PM (2002) In utero bacterial endotoxin exposure causes loss of tyrosine hydroxylase neurons in the postnatal rat midbrain. Mov Disord 17:116–124

18. Ling Z, Zhu Y, Tong C, Snyder JA, Lipton JW, Carvey PM (2006) Progressive dopamine neuron loss following supra-nigra lipopolysaccharide (LPS) infusion into rats exposed to LPS prenatally. Exp Neurol 199:499–512

19. Boksa P (2010) Effects of prenatal infection on brain development and behavior: a review of findings from animal models. Brain Behav Immun 24:881–897

20. Feleder C, Tseng KY, Calhoon GG, O'Donnell P (2010) Neonatal intrahippocampal immune challenge alters dopamine modulation of prefrontal cortical interneurons in adult rats. Biol Psychiatry 67:386–392

21. Maciag D, Simpson KL, Coppinger D, Lu YF, Wang Y, Lin RCS, Paul IA (2006) Neonatal antidepressant exposure has lasting effects on behavior and serotonin circuitry. Neuropsychopharmacology 31:47–57

22. Weaver KJ, Paul IA, Lin RCS, Simpson KL (2010) Neonatal exposure to citalopram selectively alters the expression of the serotonin transporter in the hippocampus: dose-dependent effects. Anat Rec 293:1920–1932

23. Zhang JL, Darling RD, Paul IP, Simpson KL, Chen K, Shih JC, Lin RCS (2011) Altered expression of tyrosine hydroxylase in the locus coeruleus noradrenergic system in citalopram neonatally exposed rats and monoamine oxidase A knock out mice. Anat Rec 294:1685–1697

24. Scott AL, Bortolato M, Chen K, Shih JC (2008) Novel monoamine oxidase A knock out mice with human-like spontaneous mutation. NeuroReport 19:739–743

25. Bortolato M, Godar SC, Alzghoul L, Zhang JL, Darling RD, Simpson KL, Bini V, Chen K, Wellman CL, Lin RCS, Shih JC (2012) Monoamine oxidase A and A/B knockout mice display autistic-like behavior. Int J Neuropsychopharmacol 16:869–888

26. Fan LW, Tien LT, Zheng BY, Pang Y, Lin RCS, Simpson KL, Ma TG, Rhodes PG, Cai ZW (2011) Dopaminergic neuronal injury in the adult rat brain following neonatal exposure to lipopolysaccharide and the silent neurotoxicity. Brain Behav Immun 25:286–297

27. Pang Y, Cai ZW, Rhodes PG (2005) Effect of tumor necrosis factor-alpha on developing optic nerve oligodendrocytes in culture. J Neurosci Res 80:226–234

28. Pang Y, Zheng BY, Simpson KL, Cai ZW, Rhodes PG, Lin RCS (2012) Neuron-oligodendrocyte myelination co-culture derived from embryonic rat spinal cord and cerebral cortex. Brain Behav 2:53–67

29. Matesic DF, Lin RCS (1994) Microtubule-associated protein 2 as an early indicator of ischemia-induced neurodegeneration in the gerbil forebrain. J Neurochem 63:1012–1020

30. Simpson KL, Lin RCS. 2002. Cellular and molecular mechanisms of ischemic tolerance.

New concepts in cerebral ischemia; CRC Press, New York. 2002, pp. 263–306

31. Bauman ML, Kemper TL (2005) Neuroanatomic observations of the brain in autism: a review and future directions. Int J Dev Neurosci 23:183–187

32. Meyer U, Yee BK, Feldon J (2007) The neurodevelopmental impact of prenatal infections at different times of pregnancy. Neuroscientist 13:241–256

33. Meyer U, Feldon J, Fatemi SH (2009) In vivo rodent models for the experimental investigation of prenatal immune activation effects in neurodevelopmental brain disorders. Neurosci Biobehav Rev 33:1061–1079

34. Patterson PH (2007) Maternal effects on schizophrenia risk. Science 318:576–577

35. Fan LW, Tien LT, Lin RCS, Simpson KL, Rhodes PG, Cai ZW (2011) Neonatal exposure to lipopolysaccharide enhances vulnerability of nigrostriatal dopaminergic neurons to rotenone neurotoxicity in later life. Neurobiol Dis 44:304–316

36. Cai ZW, Fan LW, Kaizaki A, Tien LT, Ma TG, Pang Y, Lin SY, Lin RCS, Simpson KL (2013) Neonatal systemic exposure to lipopolysaccharide enhances susceptibility of nigrostriatal dopaminergic neurons to rotenone neurotoxicity in later life. Dev Neurosci 35:155–171

37. Wood P, Okada E, Bunge R (1980) The use of networks of dissociated rat dorsal root ganglion neurons to induce myelination by oligodendrocytes in culture. Brain Res 196:247–252

38. Wang Z, Colognato H, Ffrench-Constant C (2007) Contrasting effects of mitogenic growth factors on myelination in neuron-oligodendrocyte co-culture. Glia 55:537–545

39. Yang Y, Lewis R, Miller RH (2011) Interactions between oligodendrocyte precursors control the onset of CNS myelination. Dev Biol 350:127–138

40. Chen Z, Ma Z, Wang Y, Li Y, Lu H, Fu S, Hang Q, Lu PH (2010) Oligodendrocyte-spinal cord explant co-culture: an in vitro model for the study of myelination. Brain Res 1309:9–18

41. Merrill J (2009) In vitro and in vivo pharmacological models to assess demyelination and remyelination. Neuropsychopharmacology 34:55–73

42. Connor JA, Razani-Borouhrtdi S, Greenwood AC, Cormier RJ, Petrozzino JJ, Lin REC (1999) Reduced voltage-dependent Ca++ signaling in CA1 neurons after brief ischemia in gerbils. J Neurophysiol 81:299–306

Neuromethods (2016) 115: 139–166
DOI 10.1007/7657_2015_77
© Springer Science+Business Media New York 2015
Published online: 20 April 2016

Using Sequential Dual-Immunogold-Silver Labeling and Electron Microscopy to Determine the Fate of Internalized G-Protein-Coupled Receptors Following Agonist Treatment

Elisabeth J. Van Bockstaele, Janet L. Kravets, Xin-Mei Wen, and Beverly A.S. Reyes

Abstract

Internalization of ligand-receptor complexes is a consequence of receptor signaling resulting in receptors being recycled to the cell surface or transported to lysosomes for eventual degradation. Resolving the fate of internalized receptors allows predictions to be made regarding changes in cellular sensitivity and the ability to test specific synaptic models of interaction between related receptor systems. Dual-labeling immunohistochemistry employing visually distinct immunoperoxidase and immunogold markers has been an effective approach for elucidating complex receptor profiles at the synapse and to definitively establish the localization of individual receptors and ligands to common cellular profiles. However, the combination of dual-immunogold-silver labeling of distinct antigens offers some unique benefits for resolving interactions between G-protein-coupled receptors (GPCR), their interacting proteins, or downstream effectors. This approach provides superior subcellular localization of the antigen of interest while preserving optimal ultrastructural morphology. Pre-embedding methods are also more appropriate than post-embedding methods for localization of immunoreactivity at extrasynaptic sites making quantification of GPCR distribution more suitable. Here, we provide detailed methodologies of the use of different sized immunogold particles to analyze the association of the mu-opioid receptor (MOR) with lysosome-associated membrane protein (LAMP), an integral membrane protein that is associated with lysosomes or early endosome antigen (EEA), a protein associated with early endosomes, following intracerebroventricular (icv) administration of saline or the opiate agonist [D-Ala2, N-Me-Phe4, Gly-ol5]-enkephalin (DAMGO) in the striatum of male Sprague-Dawley rats.

Keywords Receptor internalization, Trafficking, Ultrastructure, Interacting proteins, Endosomes

1 Introduction and Background

G-protein-coupled receptors (GPCRs) [1, 2], also called seventransmembrane receptors, are a large family of cell-surface proteins capable of binding a diverse array of molecules that mediate a myriad of physiological responses to hormones, neurotransmitters, and environmental stimulants [3]. The importance of GPCR signaling in dictating cellular sensitivity has been extensively characterized [4–10]. Dysregulation of GPCRs has been associated with a host of pathophysiological conditions [10]. Given the diverse modulatory functions attributed to GPCRs, this family of cell surface proteins has been an important target in the development of therapeutic tools for the treatment of various disease conditions [11].

Neuroanatomical techniques used to discern the cellular distribution of GPCRs have revealed important sites of action that can inform experimental approaches using pharmacological manipulation to target these systems under different experimental conditions [12–16]. However, even under carefully controlled conditions, light and fluorescence microscopy techniques provide only limited resolution. Electron microscopy provides enhanced subcellular precision. The pre-embedding immunocytochemical approach combined with electron microscopy has several advantages. This approach maintains morphological preservation while preserving discrete subcellular localization of the antigen of interest [17]. Furthermore, pre-embedding immunolabeling can be more appropriate than post-embedding labeling for localizing immunoreactivity at extra-synaptic sites and, therefore, is particularly useful for determining the regional distribution of GPCRs [18]. Post-embedding labeling, although useful for some antigens, can yield a high degree of nonspecific adhesion and, therefore, specificity of the reaction deposit is difficult to discern. The combination of peroxidase labeling, which appears as a dense homogeneous precipitate within cellular compartments, with silver-enhanced immunogold labeling can unequivocally establish coexistence of distinct receptor proteins in common cellular elements. Although the pre-embedding immunogold-silver approach may produce lower estimates of receptor number than immunoperoxidase labeling due to differences in reagent penetration [17], limitations of the experimental approach are lessened by only examining sections from the outer surface of the tissue for data analysis. Furthermore, whenever penetration is considered more important than preservation of fine structure of the neuropil, enhancement methods (such as increased detergents, i.e., Triton X-100) can be considered [17].

The immunoelectron microscopy approach can also be useful for examining trafficking of GPCRs following agonist administration [12–14, 16]. This approach offers the potential for determining cell surface versus intracellular localization, as well as the association with various identifiable cellular organelles. Most studies of receptor trafficking examine these adaptations in modified cell culture preparations rather than in vivo models. Importantly, the development of enhanced detection methods using sequential immunogold-silver labeling [19–21] allows for testing novel hypotheses regarding putative interactions of different receptor subtypes, their interacting proteins, or downstream effectors. Analysis of brain tissue sections that are dually labeled with immunogold-silver processed sequentially, such that two different sized immunogold-silver particles are produced, allows for greater spatial resolution when compared to traditional electron-dense markers such as immunoperoxidase. Obtaining different sized immunogold-silver particles is achieved by incubating with one ultrasmall gold conjugate, followed by silver enhancement, and then incubating with the second ultrasmall gold conjugate, followed by additional silver enhancement [22]. This results in two groups of silver-enhanced particles: smaller particles that are enhanced once and larger particles that are enhanced twice [19, 22, 23]. In summary, triple-labeling immunohistochemistry employing visually distinct immunoperoxidase and dual-immunogold markers can be used to elucidate complex receptor profiles at the synapse and to definitively establish the localization of individual receptors and neuromodulators to common cellular profiles.

1.1 Trafficking of GPCRs: Using MOR as a Model System

Trafficking of receptors by agonist stimulation plays a significant role in modulating effects on cellular activity [6, 24–26]. Agonist stimulation of GPCRs can result in internalization of the receptor protein into intracellular vesicles. Subsequently receptors are either recycled back to the cell surface or further transported to lysosomes for eventual degradation [27–29]. Receptor internalization pathways have been shown to include clathrin-coated vesicles and non-coated vesicles [6, 30–33]. Some MOR agonists induce a rapid and dramatic shift in receptor distribution that is readily detectable with electron microscopic analysis. Indeed, etorphine produces a depletion of plasma membrane receptors from both parasynaptic sites adjacent to axon terminals as well as receptors at extrasynaptic plasma membrane sites that are ensheathed by astrocytic processes [34]. However, neither acute nor chronic morphine treatment is associated with a significant change in receptor density or distribution [35–37]. These results lend weight to previous data using low-resolution receptor autoradiography that the neural adaptations associated with the development of opioid tolerance and dependence are not mediated by changes in the availability of receptors, but rather support the notion that these changes likely involve adaptations within receptor signaling cascades.

MORs show a ligand-dependent mechanism of desensitization and internalization [38, 39] that has been extensively described [39, 40]. Specifically, MORs internalize in response to some agonists such as the endogenous opioid peptides, methionine[5] and leucine[5]-enkephalin (ENK), and etorphine [41] but not the partial agonist, morphine [42–45]. Studies examining the pathway of internalization of GPCRs involving β-arrestins have supported a role for clathrin-coated pit-mediated endocytosis in the internalization of receptors [46]. The idea that β-arrestins target GPCRs for endocytosis via clathrin-coated vesicles has been corroborated by studies showing that β-arrestins interact directly with components of the endocytotic machinery involved in the formation of clathrin-coated pits [47, 48]. Evidently, the association of MOR with clathrin-coated vesicles is rapid as gold-silver labeling for MOR was predominantly localized to early endosomes with clathrin-coated vesicles nearby. Whereas other GPCRs, such as the A1 adenosine receptor [49, 50], internalize at slower rates ($11/2 = 90$ min), the rate at which etorphine causes internalization of MORs is quite rapid (15 min). These kinetic differences suggest that GPCR internalization may be mediated by multiple endocytotic mechanisms and/or that structural heterogeneity between receptor subtypes modulates their relative affinities to bind endocytotic adaptor proteins [6].

DAMGO treatment results in arrestin-2 translocation to the plasma membrane, which leads to MOR desensitization and significant internalization [39, 40, 51–53]. Finally, different agonists can activate distinct MOR signaling pathways beyond differences in internalization [54, 55].

1.2 Using ImmunoEM to Localize MOR-Interacting Proteins

We have recently identified Wntless (WLS) as a MOR-interacting protein that may possibly serve as a substrate underlying the alterations in neuronal structure, synaptic organization, and molecular adaptations characteristic of opioid dependence [19–21, 56]. MOR/WLS interaction may play a role in agonist-induced redistribution of WLS which is differentially influenced by opiate agonists. As binding partners, interactions between MOR and WLS may occur in the cytosolic compartment or the plasma membrane of cells [19]. Such associations have implications for receptor function. Johnson and colleagues reported that morphine-desensitized receptors are not internalized but are retained on the plasma membrane in the desensitized form as a result of different conformational changes of MORs being stabilized by different agonists that recruit different regulatory elements to the receptor [39]. Specifically, it was suggested that desensitization and tolerance to morphine are mediated largely by protein kinase C while desensitization by DAMGO is mediated by a G-protein-coupled receptor kinase [39, 57]. We hypothesize that when morphine binds with MOR, the morphine-enhanced interaction between MOR and WLS causes entrapment of WLS at the cell surface and WLS is inefficiently internalized. As a result, a larger proportion of MOR and WLS is present at the plasma membrane enabling more MOR to be available for activation by morphine. This event effectively sequesters WLS, thereby inhibiting WLS function in mediating Wnt secretion since a significant inhibition of Wnt secretion was observed in morphine-treated 293-MOR cells [19]. While WLS is inefficiently internalized after morphine, WLS is efficiently internalized in the presence of DAMGO [19].

An interaction between WLS and Wnt proteins has been demonstrated in the Golgi complex and these two proteins shuttle together to the plasma membrane where Wnt proteins are then released [58]. It is via a clathrin-mediated endocytosis that WLS is internalized, while it is via the retromer complex that WLS is recycled back to the Golgi. WLS binds with MOR and their association occurs both in the cytoplasmic compartment and on the plasma membrane [19]. Following morphine treatment, the level of MOR/WLS complexes significantly increased using immunoprecipitation and immunohistochemistry [19]. In striatal neurons, we reported that following morphine treatment, WLS is redistributed from the cytoplasmic compartment to the plasma membrane [19, 21]. It is likely that following morphine treatment, MORs are inefficiently internalized as we [34] and others [43] have previously reported. Furthermore, using locus coeruleus neurons we recently reported that morphine treatment caused a significant redistribution of WLS to the plasma membrane in addition to an increase in the proximity of gold-silver labels for MOR and WLS [56]. This event then leads to an increase in the association between WLS and MOR that may potentially cause WLS to be translocated to the plasma membrane and effectively sequestered there. In turn, WLS recycling is reduced, thereby inhibiting Wnt secretion. We have also demonstrated that MOR and WLS are co-localized in tyrosine hydroxylase-containing neurons [56].

The goal of this chapter is to provide a detailed methodological protocol for pre-embedding immunocytochemistry using sequential silver-enhanced dual-gold labeling of antigens of interest combined with electron microscopy. At the outset, the experimenter must clearly define the question to be answered so as to best fit the experimental design to the hypothesis being tested. Here, using a similar experimental design, two questions will be answered: (1) Following administration of DAMGO, does MOR internalize (e.g., shift its distribution from the plasma membrane to the cytoplasm) and associate with EEA or LAMP in the striatum when compared to control? (2) Following administration of morphine does MOR associate with EEA or LAMP in the striatum when compared to control?

2 Equipment, Materials, and Setup

Immunoelectron microscopy using sequential silver-enhanced dual-gold labeling entails a multiple-step process. The first step involves transcardial perfusion of the animal for brain extraction. The second step involves postfixation and sectioning of the brain. Pre-embedding dual-immunogold labeling is then initiated followed by embedding, ultrathin sectioning, and counterstaining. Samples are then ready for analysis using a transmission electron microscope. All essential equipment and materials necessary to conduct pre-embedding sequential silver-enhanced dual-gold labeling of antigens of interest combined with electron microscopy are listed as follows:

Personal protective and safety equipment	
Gloves (BioExpress)	Biohazard bags (Fisher Scientific)
Masks (Fisher Scientific)	Chemical waste containers
Lab coat	Eye wash station

Transcardial perfusion of the animal is a crucial step in achieving optimal preservation of ultrastructural morphology with clearly

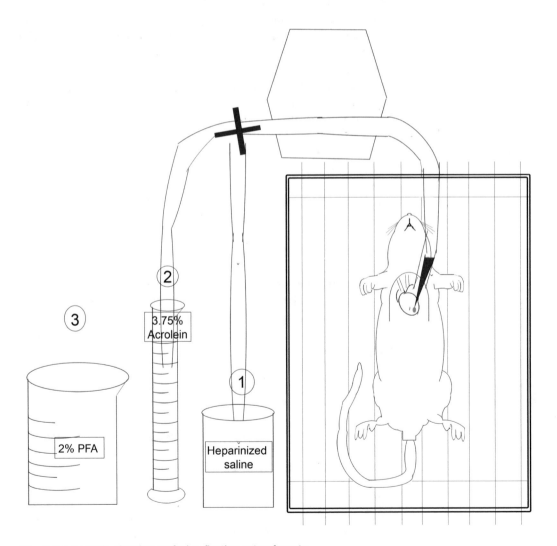

Fig. 1 A schematic showing perfusion fixation setup for rats

apparent immunocytochemical labeling. Figure 1 shows a schematic illustration of a perfusion fixation setup.

Materials and solutions for perfusion fixation	
Equipment	Balance (Ohaus Adventurer)
Glass beakers (Fisher Scientific)	Silicon tubing (Cole-Parmer Instrument Co.)
Graduated cylinder (Nalgene)	Spatula (VWR)
Magnetic stir bar (Fisher Scientific)	Erlenmeyer 500 mL filtration flask (VWR)
Hot plate (Fisher Scientific)	Buchner funnel (Coors)
Thermometer (Fisher Scientific)	*Solutions*
Tray (New Pig)	Sodium phosphate monobasic (Sigma)
Enclosed pan (New Pig)	Sodium phosphate dibasic (Sigma)
Syringe needle (BD Falcon)	Paraformaldehyde (Electron Microscopy Services)
Parafilm (American National Can)	Deionized water
Scissors (FST)	Sodium hydroxide (Fisher Scientific)
Scalpel (FST)	Isoflurane (Southmedic Inc)
20 mL glass vial (Fisher Scientific)	Sodium pentobarbital (Sigma)
Whatman #3 Filter paper (Sigma)	Acrolein (Polysciences)
Masterflex perfusion pump (Cole-Parmer Instrument Co.)	Normal saline with heparin (Henry Schein)
18 gauge stainless steel tubing (Small Parts Inc.)	

Following the transcardial perfusion, the brain is collected and sectioned with a vibrating microtome (Fig. 2) prior to processing for sequential dual-immunogold histochemistry.

Materials and solutions for vibratome sectioning the brain	
Equipment	Fine forceps (Electron Microscopy Services)
Vibrating microtome (Leica Microsystems)	Specimen block (Leica Microsystems)
Brain blocking mold (Ted Pella Inc)	*Solutions*
Parafilm (American National Can)	Ethylene glycol (Sigma)
Razor blades (Personna American Safety Razor Co.)	Sucrose (Fisher Scientific)
12-Well tissue culture dishes (VWR)	Agar (Sigma)

(continued)

(continued)

Materials and solutions for vibratome sectioning the brain	
Camel and sable hairbrushes (Ted Pella Inc)	Super Glue (Electron Microscopy Services)
Microwave (Sanyo)	Sodium phosphate monobasic (Sigma)
Glass beaker (Fisher Scientific)	Sodium phosphate dibasic (Sigma)

The sequential dual-immunogold labeling generates two different groups of silver-enhanced particles: smaller particles that are enhanced once and larger particles that are enhanced twice (Fig. 2).

Materials and solutions for sequential dual-immunogold labeling	
Equipment	0.1 M Phosphate buffer
Glass petri dishes (Corning Life Sciences)	0.1 M Tris-buffered saline (TBS)
Camel and sable hairbrushes (Ted Pella Inc)	0.5 % Bovine serum albumin (Sigma)
Pipettes (Eppendorf)	Washing incubation buffer
Glass petri dishes (VWR)	0.2 M Citrate buffer
6-Well tissue culture dishes (VWR)	0.01 M Phosphate-buffered saline (PBS)
Coors dishes (Coors)	Primary antibodies
Aclar fluorhalocarbon film (Electron Microscopy Services)	Ultrasmall gold-conjugated IgG (Electron Microscopy Services)
Rotator (Lab-Line)	Enhancement condition solution (Aurion)
Thermolyne, Roto-mix (Sigma)	R-Gent Se-EM enhancement mixture (Electron Microscopy Services)
Accument Basic AB15 pH meter (Fisher Scientific)	2 % Glutaraldehyde (Electron Microscopy Services)
Oven (Fisher Scientific)	2 % osmium tetroxide (Electron Microscopy Services)
20 mL glass vials (Fisher Scientific)	100 % Ethanol (Decon Labs Inc)
Wooden applicator sticks (VWR)	Propylene oxide (Electron Microscopy Services)
Pasteur pipettes (VWR)	Epon
Fume hood (Flow sciences)	EM-bed 812 (Electron Microscopy Services)

(continued)

(continued)

Materials and solutions for sequential dual-immunogold labeling	
Solutions	Dodecenyl succinate anhydride (Electron Microscopy Services)
Sodium borohydride (Sigma)	NADIC^R methyl anhydride (Electron Microscopy Services)
Sodium phosphate monobasic (Sigma)	2,4,6-(Tri(dimethylamino ethyl phenol)) (Electron Microscopy Services)
Sodium phosphate dibasic (Sigma)	Coldwater fish gelatin (Electron Microscopy Services)

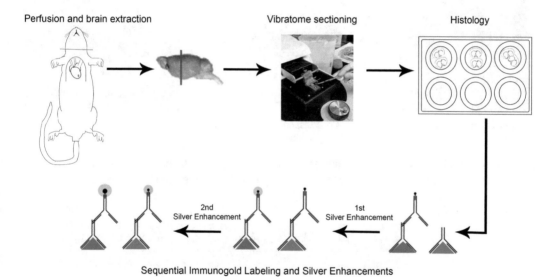

Fig. 2 Sectioning the brain on the vibrating slicer, sequential-immunogold labeling, and silver enhancement

Once the tissue sections are processed for sequential dual-immunogold labeling, they are flat embedded on a film of aclar fluorhalocarbon and mounted on a polymerized block using glue and the block face is trimmed (ensuring that the block contains the region of interest) in the shape of a trapezoid. Trimming the block face is critical for obtaining ribbons of tissue during the ultramicrotome-sectioning process (Fig. 3).

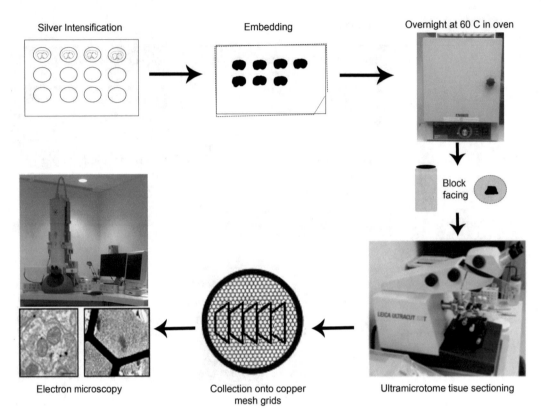

Fig. 3 A schematic diagram showing procedures involved in electron microscopy starting from silver enhancement up to visualizing an ultrathin section using a transmission electron microscope

Materials for ultramicrotome tissue sectioning and electron microscopy	
Ultramicrotome (Leica Microsystems)	Fine forceps (Electron Microscopy Services)
Diamond knife (DiATOME)	Electron microscope (Morgagni, Fei Company)
Copper mesh grids (Electron Microscopy Services)	AMT advantage HR/HR-B CCD camera system (Advanced Microscopy Techniques Corp.)
Hair curl (Electron Microscopy Services)	Adobe Photoshop CS4 software (Adobe Systems Inc)

Tissue sections mounted onto the copper grids are counterstained prior to examination using the electron microscope. This step enhances contrast in the tissue by exposure to electron-dense, heavy metal stains.

Materials for counterstaining the tissue sections	
Fine forceps (Electron Microscopy Services)	Dental wax (Electron Microscopy Services)
5 mL syringe (BD Syringe)	Uranyl acetate (Electron Microscopy Services)
Leur lock filter tip (BD Syringe)	Deionized water
Glass petri dish (Corning Life Sciences)	Lead nitrate (Electron Microscopy Services)
Sodium hydroxide pellets (Fisher Scientific)	Sodium hydroxide pellets (Fisher Scientific)
10 mL glass beakers (Fisher Scientific)	Lead citrate (Electron Microscopy Services)
Aluminum foil (Fisher Scientific)	

Listed below are the solutions that are needed to conduct the sequential dual-immunogold labeling:

Solutions:

TBS pH 7.6	Washing incubation buffer
4 L Deionized water	500 mL 0.01 M PBS
48.4 g Trizma base	2.5 mL Gelatin stock
36 g Sodium chloride	1.0 g BSA
0.5 % Bovine serum albumin (BSA)	*0.2 M Citrate buffer*
100 mL 0.1 M TBS	500 mL Deionized water
0.5 g BSA	29.45 g Sodium citrate
Antibody incubation buffer	*Epon*
0.1 g BSA	24 mL EM-bed 812
100 mL 0.1 M TBS	15 mL DDSA
0.01 M Phosphate-buffered saline (PBS) pH 7.40	13.5 mL NMA
1 L Deionized water	525 µL DMP-30
1.09 g Sodium phosphate dibasic	
0.32 g Sodium phosphate monobasic	
9.0 g Sodium chloride	

3 Procedures

3.1 General Comments

A significant consideration in any immunocytochemical approach involves ensuring specificity of the antisera used in the experiment. An antiserum is a preparation containing the immune globulins from the serum of an animal that has been immunized with a specific antigen. Only a small portion of the immune globulins in that serum will be directed against the antigen used for immunization, hence the need for conducting sufficient control experiments to determine that the epitope that is recognized in the tissue belongs to the original antigen. A useful resource for outlining important and stringent control experiments (as well as discussion of monocolonal, polyclonal, and affinity-purified antisera) can be found in an editorial by Saper and Sawchenko [59]. Briefly, to authenticate a staining pattern, an adsorption control is required. This involves mixing the antiserum with an excess of the immunizing peptide, exposing this cocktail to fixed brain tissue, and confirming that no staining is achieved. A Western blot of the tissue of interest (as well as a negative control) should be conducted to demonstrate that the antibody recognizes the protein of the appropriate molecular weight. Another important control experiment is to conduct staining for the antigen of interest in a genetically engineered mouse where the antigen has been eliminated and compare the staining pattern to a wild-type mouse. Finally, control experiments should involve omission of the primary antisera to test the specificity of secondary antibodies. In general, although some antibodies may have been extensively characterized in other laboratories, best practices include erring on the side of using more controls than necessary to unequivocally establish the specificity of antisera used in immunocytochemical experiments.

The following is an example of a narrative used to describe the specificity of a MOR-interacting protein, defined as WLS, whose characterization and specificity were previously described [19, 60]. The WLS antibody was raised in chicken against a peptide antigen corresponding to the C-terminal 18 amino acids (HVDGPTEIYKLTRKEAQE) of human WLS (Gene-Tel Laboratories, Madison, WI), which is identical to the rat and mouse peptide sequence. A polyclonal antibody, WLS of IgY subtype was harvested from yolk and affinity-purified prior to use [19, 60]. The specificity of WLS was ascertained by transfecting a FLAG/6x His-tagged WLS cDNA into HEK293 cells. Using Western blot analysis WLS immunoreactivity was expressed in lysates prepared from transfected cells, which were first probed with an anti-FLAG antibody, then stripped, and reprobed with anti-WLS antibody. A band migrating with the identical molecular mass was detected with anti-FLAG and anti-WLS antibodies indicating that WLS antibody was specific for the WLS protein. In transfected cells, anti-WLS antibody recognized a single-protein species with an approximate molecular weight of about 50 kDa [60]. Furthermore, preabsorption of WLS with its antigenic peptide did not reveal any staining for WLS immunoreactivity in the rat frontal cortex. Due to the fact that WLS knockouts are embryologically lethal, staining in WLS-knockout mice is not possible.

3.2 Transcardial Perfusion, Tissue Sectioning

Rats or mice should be housed three to a cage (20 °C, 12-h light/ dark cycle, lights on 7:00 a.m.) with food and water freely available. All care and use of animal procedures should be approved by an Institutional Animal Care and Use Committee and follow the revised *Guide for the Care and Use of Laboratory Animals* (1996), the Health Research Extension Act (1985), and the PHS Policy on Humane Care and Use of Laboratory Animals (1986). Rats and mice should be anesthetized to the point of insensitivity to physical or emotional distress, as determined by tail pinch and loud noise. Once the experimenter confirms insensitivity to pain, transcardial perfusion should proceed. Kravets and colleagues [61] and Milner and colleagues [62] recently published detailed perfusion fixation procedures for rats and mice, respectively, that are recommended for the experimental approach provided here. These references supply helpful illustrations of the perfusion setup and the required steps to achieve a successful perfusion fixation. Briefly, rats were deeply anesthetized with sodium pentobarbital (70 mg/kg; [14]) and perfused transcardially through the ascending aorta with 50 mL of 3.8 % acrolein and 200 mL of 2 % paraformaldehyde in 0.1 M phosphate buffer (PB; pH 7.4). Immediately following perfusion–fixation, the brains were removed, cut into 1–3 mm coronal slices, and placed in the same fixative for an additional 30 min. Forty-micrometer-thick sections were cut through the rostrocaudal extent of the striatum using a Vibratome and collected into 0.1 M PB.

The immunoelectron microscopy procedures from the first day through the fifth day are step by step described as follows:

Day 1
(Approximately 2.5 h)

1. 1 % Sodium borohydride for 30 min.

   – 1 g Sodium borohydride in 100 mL 0.1 M PB (pH 7.3–7.4).

   *Note: In preparing 0.1 M PB, do not use HCl acid to adjust the pH; rather weigh appropriate amount of sodium monobasic and dibasic to arrive at 7.3–7.4 pH. The appearance of holes in tissues happens when a buffer contacting saline/chloride in the solution comes in contact with intensified silver.

2. Wash 3× in 0.1 M PB.

3. Wash in 0.1 M TBS (pH 7.6).

   2× at 5 min each wash.

4. 0.5 % BSA in 0.1 M TBS.

   30 min.

0.5 % BSA	0.1 M TBS
0.5 g	100 mL
0.1 g	20 mL

5. Wash in 0.1 M TBS.

   2× at 5 min each wash.

6. Primary antibody incubation.

0.1 % BSA	0.1 M TBS
0.1 g	100 mL
0.02 g	20 mL

Incubate the primary antibody overnight at room temperature (1 mL/tube).

*Note: Triton X should not be added in the incubation solution for tissues intended for electron microscopy.

**Primary antibody dilutions

MOR (rabbit) 1:2,500 dilution and LAMP (mouse) 1:500 dilution.

MOR (rabbit) 1:2,500 dilution and EEA (mouse) 1:500 dilution.

Prepare the following solutions for the next day's processing:

(A) 0.01 M PBS.

Deionized water (L)	Dibasic PO$_4$ (g)	Monobasic PO$_4$ (g)	NaCl (g)
4	4.36	1.28	36.0
2	2.18	0.64	18.0
1	1.09	0.32	9.0
0.5	0.545	0.16	4.5

pH 7.4.

(B) Washing incubation buffer.

0.01 M PBS (mL)	Gelatin stock (0.1 %) (mL)	BSA (0.2 %) (g)
500	2.5	1.0
250	1.25	0.5
125	0.625	0.25
100	0.50	0.2

pH 7.4.

*Note: If blocking with fish gelatin start thawing/melting it right away; can use water bath.

(C) 0.2 M Citrate buffer.

Deionized water (mL)	Sodium citrate (g)
500	29.45
250	14.73
200	11.78
100	5.89

pH 7.4 (use citric acid to adjust the pH).
Washes—Takes between 50 and 120 mL per 6-well dish.
Number of washes using each solution after day 1:
PB 0.1 M—16.
Tris—3.
PBS 0.01 M—8.
Washing buffer—8.
Citrate buffer—1.
PB 0.2 M—1.

Day 2
(Approximately 10–11 h)

1. Wash in 0.1 M Tris.

   3× 10 min each wash.
2. 2× Wash in 0.1 M PB.

3. Wash in 0.01 M PBS once.

4. Block in washing incubation buffer for 10 min.

5. Prepare the first ultrasmall gold-conjugated IgG (1:50) in washing incubation buffer.

   Incubate the tissues for 8 h at room temperature.
   *Note: Whichever goat anti- (rabbit or mouse, etc.) is used to match primary antibody will now be the larger particle at the end. For this experiment, we used goat anti-rabbit because we label MOR as large immunogold-silver particles.
6. Wash in washing incubation buffer for 5 min.

7. Wash in 0.01 M PBS 6× at 5 min each.

8. Wash in 0.1 M PB 2× at 5 min each (store overnight at 0.1 M PB at 4 °C).

(D) Additional materials needed for day 4:

   1. Aurion ECS (enhancement condition solution) (10× solution) stored at room temperature.

   2. R-Gent Se-EM enhancement mixture stored at 4 °C.

Day 3
(Approximately 12 h)

1. Wash in 0.1 M PB 2× at 5 min each.
2. Pre-enhancement washing: ECS, 4× at 5 min each.
   - Dilute 10× ECS solution to 1× with diH$_2$O; add ~500 μL per mini well in the 4 × 6 plates. (ECS for 12 wells: need 6 mL ~>7 mL/wash x 4 = 28 mL, therefore 2.8 mL of concentrated ECS into 25.2 mL of diH$_2$O.)

   *Note for steps 2 and 3: You may fill all the needed 4 × 6 mini wells with the respective wash solution first (rows 1–4 with 1× ECS solution and row 5 silver enhancement solution) and just move the tissues to each of the five rows below it for each wash step instead of pipetting out solution and adding new.

3. First silver enhancement: Use R-Gent SE-EM enhancement mixture for 90 min.
   - *How to prepare the developer*: Put 40 drops of activator (kept in the refrigerator) and 1 drop of initiator (kept in the freezer) into the developer bottle. Mix the developer well on a vortex.
   - *To prepare the enhancement mixture*: Put 20 drops enhancer and 1 drop developer in a well and mix in the shaker. Put 0.3–0.5 mL in each well (~5 tissues).

   *Note: Use approximately 300–500 μL per mini well (EX: *Since each drop is about 50 uL then each combo of 20 drops enhancer + 1 drop of developer = 1,050 μL.* For 12 wells you need between 4 and 6× this 21 drop combo depending on how many tissues you have per well (realizing you cannot really get half drops easily without pipetting) [12 × 300 μL = 3,600 μL/1,050 μL = 3.4× drop combo AND 12 × 500 μL = 6,000 μL/1,050 μL = 5.7× drop combo]. If you choose 5× the drop combo: Mix 100 drops of enhancer with 5 drops of developer to get 5,250 μL enhancement solution or 437.5 μL theoretically for each of the 12 wells, then you can actually pipette ~420 μL into each well safely given drop variation, etc.
   **Note: Use wood sticks to transport tissues from this stage because the paintbrushes react/affect/get clogged up with the silver enhancement.

4. Wash in citrate buffer.
5. Wash in 0.1 M PB 2× at 5 min each.
6. Wash in 0.01 M PBS once.
7. Prepare the second ultrasmall gold-conjugated IgG (1:50) in washing incubation buffer.

   Incubate the tissues for 8 h.

*Note: Antibody here ultimately corresponds to the smaller particle on EM. For this experiment we used mouse because we labeled EEA or LAMP as small immunogold-silver particles.

8. Wash in washing incubation buffer 6× at 5 min each.

9. Wash in 0.1 M PB 2× at 5 min each (store overnight at 0.1 M PB at 4 °C).

Day 4
(Approximately 5–5.5 h)

1. Wash in 0.1 M PB 2× at 5 min each.

2. Incubate tissue sections at 2 % glutaraldehyde (EM grade), 10 min in hood.

    – To prepare 2 % glutaraldehyde: Mix 10 mL of 25 % glutaraldehyde in 40 mL 0.01 M PBS.

3. Wash in 0.1 M PB 2× at 5 min each.

4. Wash in distilled water 4× at 5 min each.

    *Note: Can set up 4 × 6 dish the same way (as day 3) for diH$_2$O (four rows) and one row for silver enhancement.

5. Second silver enhancement: Use R-Gent SE-EM enhancement mixture for 60 min.

    How to prepare the developer: Put 40 drops of activator (kept in the refrigerator) and 1 drop of initiator (kept in the freezer) into the developer bottle. Mix the developer well on a vortex.
    To prepare the enhancement mixture: Put 20 drops enhancer and 1 drop developer in a well and mix in the shaker. Put 0.3–0.5 mL in each well (~5 tissues).
    *Note: See day 3 suggestions for calculating total number of 20:1 drops for your # of wells.

6. Place the tissue sections in Coors dishes containing 0.1 M PB.

    Take one tissue section and mount on glass slide (optional).

7. Remove 0.1 M PB and replace with 2 % osmium tetroxide. Incubate for 1 h.

    – To prepare 2 % osmium tetroxide: 10 mL 4 % osmium tetroxide + 10 mL 0.2 M PB.

8. Remove the 2 % osmium tetroxide and be sure to place it in a biohazard bottle container. Replace with 0.1 M PB.

9. Dehydration steps: Dehydrate in ascending concentration of alcohol at 30, 50, 70, and 95 % for 5 min; 100 % for 10 min each (2×).

    Propylene oxide for 10 min each (2×).

    *Note: Take care not to let tissues fold at this stage because they become more brittle and are harder to unfold later without breaking.

10. Replace propylene oxide with a 1:1 solution of Epon and propylene oxide and incubate overnight.

Day 5
(Approximately 3 h)

To prepare 100 % Epon

Total volume (mL)	EM-bed 812 (mL)	DDSA (mL)	NMA (mL)	DMP-30 (μL)
52.5	24	15	13.5	525
39.38	18	11.25	10.13	393
26.25	12	7.5	6.75	262
13.13	6.0	3.75	3.38	131

1. Rotate the tissue sections in 100 % Epon for 2 h.

    *Note: When moving tissues from overnight into weigh boats into 100 % Epon if tissues stick to bottom/side of jar keep rotating it in your hands to gently resuspend tissues DO NOT scrape them off the sides with the wood stick.

2. Flat embed between two sheets of aclar fluorhalocarbon film.

    *Note: Make one sheet of aclar smaller than the other to easily pull apart later. Unfold tissues in weigh boats where that Epon-containing tube is dumped. Place all desired tissues on bottom aclar, dab/blot excess Epon around tissues off (with Kim wipe or cotton swab), and then lay smaller aclar on top pushing all air bubbles off and away from your tissues.

3. Bake the aclar in the oven (55–60 °C) overnight.

Control sections for each experiment were run in parallel in which the primary antisera were omitted but the rest of the processing procedure was identical. Sections processed in the absence of primary antibodies did not exhibit any immunoreactivity. Some sections were also processed without the secondary antibodies but the rest of the processing procedure was identical. Sections processed in the absence of the secondary antibodies did not exhibit any immunoreactivity. In some sections one of the primary antisera was omitted but the rest of the processing procedure was identical. The specific primary antiserum omitted did not show any immunoreactivity on tissue sections examined. Sections were collected on copper mesh grids and examined using an electron microscope (Morgagni, Fei Company, Hillsboro, OR, USA). Digital images were viewed and captured using the AMT advantage HR/HR-B CCD camera system (Advance Microscopy Techniques Corp., Danvers, MA, USA). Figures were assembled and adjusted for brightness and contrast using Adobe Photoshop.

3.3 Data Analysis

To determine whether levels of spurious silver grains could contribute to false positives, blood vessels and myelinated axons (structures that should not contain MOR immunolabeling) were counted in random ultrathin sections. Minimal spurious labeling was identified. Therefore, the criteria for defining a process as immunolabeled was by using detection of at least 2–3 silver grains in a cellular profile. For quantification of MOR and EEA or LAMP, tissue sections from three rats of each group (saline, morphine, and DAMGO) with optimal preservation of ultrastructural morphology were used. Only tissue sections that showed dual labeling for MOR and LAMP or MOR and EEA were used for the electron microscopic analysis. Using bright-field microscopy (Fig. 4) and electron microscopy (Fig. 5), the distribution of EEA and LAMP immunoreactivities is shown in the locus coeruleus. The identification of cellular elements was based on the standard morphological criteria [63, 64]. Dendrites usually contained endoplasmic reticulum and were postsynaptic to axon terminals. Axon terminals contained synaptic vesicles and were at least 0.3 mm in diameter. A varicosity

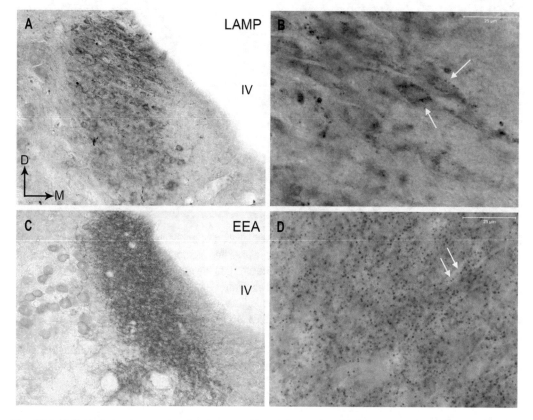

Fig. 4 Bright-field photomicrographs showing peroxidase labeling for LAMP (Panels **a** and **b**) and EEA (Panels **c** and **d**) in the locus coeruleus. Panels **b** and **d** show higher magnification images of LAMP (Panel **b**) and EEA (Panel **d**). LAMP and EEA immunolabeling (*arrows*) is visible in the core of the locus coeruleus. *Arrows* indicate dorsal (D) and medial (M) orientation of the tissue section. *IV* fourth ventricle. Scale bars, 25 μm

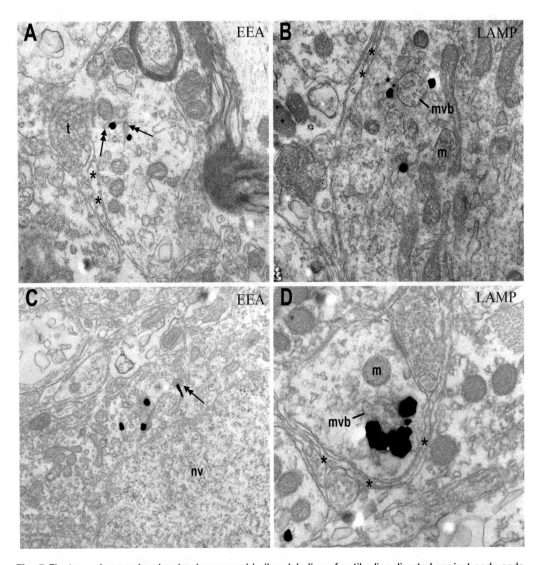

Fig. 5 Electron micrographs showing immunogold-silver labeling of antibodies directed against early endosome antigen 1 (EEA1) in panels **a** and **b**, and lysosome-associated membrane protein (LAMP) in panels **c** and **d**. (**a**) *Double-headed arrows* indicate endosome-like vesicles that are associated with gold-silver deposits for EEA-1 in a dendrite in the locus coeruleus (LC). The dendrite is apposed by an axon terminal (t) and an astrocytic process (*asterisks*). (**b**) Gold-silver deposits indicative of EEA1 are present adjacent to a multivesicular body (mvb) that contains several distinguishable spherical vesicles in an LC dendrite. Several mitochondria (m) can be discerned within the dendrite that is apposed by an astrocytic process (*asterisks*). (**c**) An LC perikarya (identified by the presence of a nucleus, nu) contains gold-silver labeling indicative of LAMP that is associated with endosome-like vesicles and an organelle that resembles a multivesicular body (*arrow*). (**d**) An LC dendrite contains gold-silver labeling for LAMP that surrounds a multivesicular body. The dendrite is enveloped by an astrocytic process (*asterisks*) and is targeted by an excitatory-type axon terminal

was considered as synaptic when it showed a junctional complex, a restricted zone of apposed parallel membranes with slight enlargement of the intercellular space, and/or associated postsynaptic thickening.

Dendrites were sampled from at least ten grids per animal containing 5–8 thin sections each that were collected from at least two plastic-embedded sections of the striatum from each animal. The quantification of MOR and LAMP or EEA-immunolabeled profiles was carried out at the plastic-tissue interface to ensure that immunolabeling was detectable in all sections used for analysis [65]. Localization of MOR, LAMP, or EEA was identified in dendritic profiles analyzed. If the immunogold-silver were associated with the plasma membrane they were classified as plasmalemmal and if the immunogold-silver were not in contact with the plasma membrane they were classified as cytoplasmic. MOR immunolabeling was identified either cytoplasmic or plasmalemmal as previously reported [34, 56, 66]. EEA and LAMP were also identified along the plasma membrane and within the cytoplasmic compartment. Semiquantitative analysis was carried out by randomly obtaining the ratio of cytoplasmic to total immunogold-silver particles for MOR and LAMP or EEA in a dendritic profile exhibiting both MOR and LAMP or EEA immunoreactivities. Labeling for each protein of interest was differentiated by electron-dense particle size and this resulted in two groups of silver-enhanced particles: smaller particles that were enhanced once and larger particles that were enhanced twice [19–22, 56, 61]. When the distribution of randomly selected gold-silver particles was plotted, a clear segregation in differently sized gold particles was revealed in a very small overlap. Dual-immunogold-silver particles were readily discernible from one another with MOR as large (cross-sectional diameter 4–44 nm) and LAMP or EEA as small (52–122 nm) gold-silver particles. Considering the importance of homogeneity in particle size for our analysis, we sought to apply certain criteria in our analysis so that the larger the size difference between the two groups of particles, the lower the likelihood of cross-reactivity. Therefore, immunogold-silver particles with a diameter ranging between 45 and 51 nm were excluded from the analysis as this represented a range where the confidence level was not high for discriminating one label from the other.

The data on the ratio of cytoplasmic to total immunogold-silver particles were quantified and analyzed. In addition, as with previous studies from our group [13, 14, 21, 34, 56, 67, 68], care was taken to ensure that control and experimental groups contained similarly sized profiles. All data were expressed as mean ± SEM. Statistical analysis was done with GraphPad Prism 5 statistical software (GraphPad Software, Inc. 2012; La Jolla, CA, USA).

Statistical analysis of the number of profiles obtained showed no significant difference in the number of dendritic profiles examined between groups. Dual-immunogold-silver labeling (large and small gold-silver particles, for MOR and EEA, respectively) that was localized in the same tissue section was readily distinguishable from each other (Fig. 6). Following morphine, large black punctate silver-enhanced gold particles, indicative of MOR, appeared on the

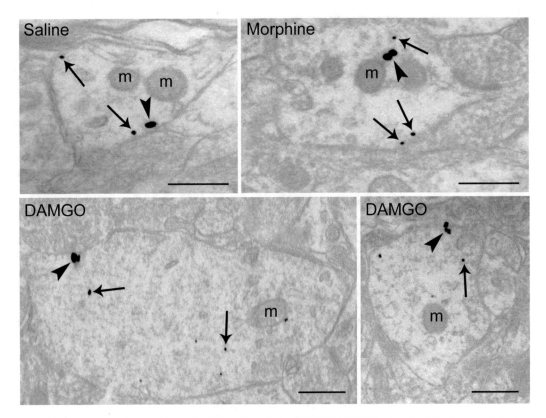

Fig. 6 Electron microscopic evidence for interactions between MOR and EEA using dual-immunogold electron microscopy. *Saline-morphine.* Sections from saline- and morphine-treated rats show small gold-silver grains for MOR (*arrows*) and large gold-silver grains for EEA (*arrowheads*) in a common striatal dendrite. MOR and EEA immunolabeling is localized along the plasma membrane and within the cytoplasm of the striatal dendrite. *m* mitochondria. Scale bars, 0.5 μm. Electron photomicrographs showing sections from rats following DAMGO treatment. *DAMGO.* Following DAMGO treatment, MOR immunolabeling (*arrows*) and EEA (*arrowheads*) show a shift in distribution from the plasma membrane to the cytoplasm. *m* mitochondria. Scale bar = 0.5 μm

plasma membrane (Fig. 6) (0.38 ± 0.03) similar to the MOR distribution in saline-treated rats (0.36 ± 0.03). Following DAMGO, MOR shifted from the plasma membrane to the cytoplasm (0.80 ± 0.04) compared to saline and morphine ($P < 0.05$; Fig. 7). In addition, the ratio of cytoplasmic to total EEA-immunogold silver particles was significantly higher ($P < 0.05$; Fig. 7) compared to saline and morphine. Similar to dual-MOR and -EEA immunolabeled profiles, dual-immunogold-silver labelings (large and small gold-silver particles, for MOR and LAMP, respectively) were localized in the same tissue section and they were readily distinguishable from each other (Fig. 7). Consistently, MOR appeared on the plasma membrane (Fig. 8) following morphine (0.35 ± 0.05; Fig. 7) similar to the MOR distribution in saline-treated rats (0.37 ± 0.04; Fig. 6). Interestingly, the ratio of cytoplasmic to total LAMP immunogold-silver particles was significantly different between groups ($P < 0.05$).

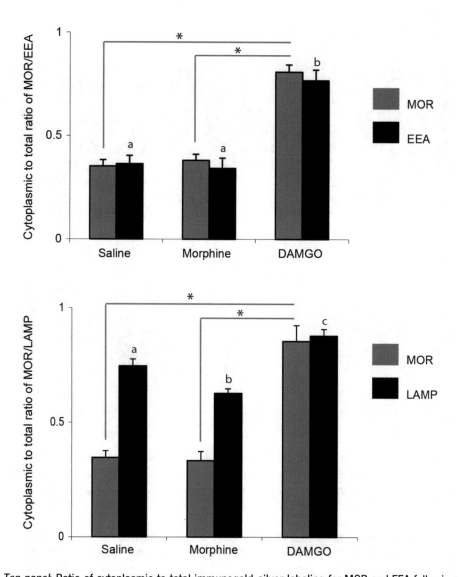

Fig. 7 *Top panel*: Ratio of cytoplasmic to total immunogold-silver labeling for MOR and EEA following opiate agonist treatment. Differential trafficking of MOR is observed in rat striatal dendrites from vehicle-treated , morphine-treated, and DAMGO-treated rats using dual-immunogold-silver for MOR (large gold-silver grains) and EEA (small gold-silver grains). DAMGO treatment caused a significant ($P < 0.05$) shift in MOR from the plasma membrane to the cytoplasmic compartment compared to saline- and morphine-treated rats. MOR remains on the plasma membrane in morphine-treated rats similar to control. DAMGO treatment significantly increased ($P < 0.05$) the ratio of cytoplasmic to total EEA immunogold-silver labeling compared to control and morphine treatment. Values are mean ± SEM of three rats per group. Values with *asterisks* are significantly different (*$P < 0.05$) from each other. Values with *different letters* are significantly different from each other. Tukey's multiple comparison tests after ANOVA. *Bottom panel*: Ratio of cytoplasmic to total immunogold-silver labeling for MOR and LAMP following opiate agonist treatment. Differential trafficking of MOR is observed in rat striatal dendrites from vehicle-treated, morphine-treated, and DAMGO-treated rats using dual-immuno-gold-silver for MOR (large gold-silver grains) and LAMP (small gold-silver grains). DAMGO treatment caused a significant ($P < 0.05$) shift in MOR from the plasma membrane to the cytoplasmic compartment compared to saline- and morphine-treated rats. MOR remains on the plasma membrane in morphine-treated rats similar to control. Morphine and DAMGO treatment significantly shifted ($P < 0.05$) the ratio of cytoplasmic to total LAMP immunogold-silver labeling compared to control. Values are mean ± SEM of three rats per group. Values with *asterisks* are significantly different (*$P < 0.05$) from each other. Values with *different letters* are significantly different from each other. Tukey's multiple comparison tests after ANOVA

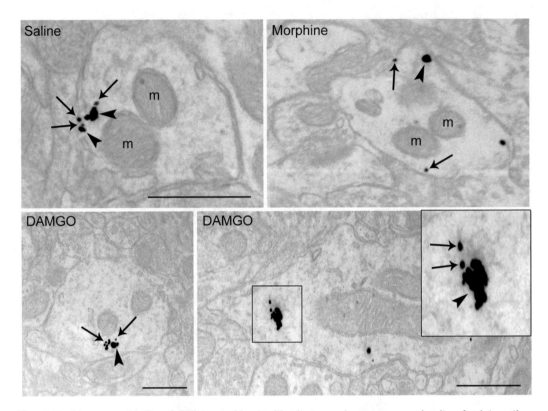

Fig. 8 Dual-immunogold-silver labeling combined with electron microscopy reveals sites for interactions between MOR and LAMP. *Saline-morphine*. Sections from saline- and morphine-treated rats indicate small gold-silver deposits for MOR (*arrows*) and large gold-silver grains for LAMP (*arrowheads*) in a common striatal dendrite. MOR and LAMP immunolabeling is localized along the plasma membrane and within the cytoplasm of a striatal dendrite. m: mitochondria. Scale bars, 0.5 μm. Electron photomicrographs showing sections from rats following DAMGO treatment. *DAMGO*. When compared to vehicle, immunogold-silver labeling (*arrows*) for MOR shows a redistribution from the plasma membrane to the cytoplasm following DAMGO administration, while immunogold-silver labeling (*arrowheads*) for EEA shows a homogeneous distribution within the cytoplasm. Scale bar = 0.5 μm

3.4 Methodological Considerations

All antisera that will be used in every research study should be well characterized and are specific to the antigen of interest. The guidelines and standards that are necessary in describing an antibody for use in neuroscience are clearly articulated in multiple publications [69–71]. Briefly, (1) a precise and complete information of the antibody (host species, nature of the antigen, and its preparation) should be provided; (2) provenance and characterization of the antibodies: often this is included in the technical information provided by the manufacturer and can also be found in previous published reports. However, Saper [71] highlighted that simply stating that the antibody has been previously characterized is not sufficient rather a reasonably critical reviewer with information on what characterization was done and what it showed. (3) Necessary controls for immunostaining: provide highest level of control that can give an assurance that the antibody is staining what it is intended to stain. This standard can be achieved by using knockout mouse or other animal, preadsorption of the antibody against the peptide. In addition, it is important that description of the control experiments performed to test the specificity of the staining pattern observed. For this purpose, (1) tissue section that does not contain the antigen of interest should be tested in parallel conditions with the tissue sections being studied; (2) another primary antibody directed against a different epitope should also be processed. As Saper and Fritschy [69, 71] emphasized controls by omission of primary antibodies provide no definitive information regarding the specificity of a primary antibody.

A potential limitation in all ultrastructural studies associated with the pre-embedding immunolabeling technique is the penetration of primary and secondary immunoreagents in thick Vibratome sections [17]. To minimize this shortcoming, analysis of ultrathin sections is carried out exclusively at the tissue/plastic interface where penetration is maximal. Second, immunolabeling of antigens is routinely reversed to ensure that patterns of immunolabeling are not attributed to differences in the detection methods of reaction products. Finally, EM data is most informative when included as part of a larger series of studies such as investigation of GPCR association with its G-protein through co-immunoprecipitation assays, cAMP assays, or GTP gamma autoradiography assays.

Acknowledgements

This work was supported by PHS grants DA09082 and DA020129. We acknowledge the experimental contributions of Dr. Jillian L. Scavone.

References

1. Hillhouse EW, Grammatopoulos DK (2006) The molecular mechanisms underlying the regulation of the biological activity of corticotropin-releasing hormone receptors: implications for physiology and pathophysiology. Endocr Rev 27(3):260–286

2. Perrin MH, Vale WW (1999) Corticotropin releasing factor receptors and their ligand family. Ann N Y Acad Sci 885:312–328

3. Rosenbaum DM, Rasmussen SG, Kobilka BK (2009) The structure and function of G-protein-coupled receptors. Nature 459 (7245):356–363

4. Bhattacharya M, Babwah AV, Ferguson SS (2004) Small GTP-binding protein-coupled receptors. Biochem Soc Trans 32(Pt 6): 1040–1044

5. Bockaert J et al (2010) GPCR interacting proteins (GIPs) in the nervous system: roles in physiology and pathologies. Annu Rev Pharmacol Toxicol 50:89–109

6. Ferguson SS (2001) Evolving concepts in G protein-coupled receptor endocytosis: the role in receptor desensitization and signaling. Pharmacol Rev 53(1):1–24

7. Ferguson SS (2007) Phosphorylation-independent attenuation of GPCR signalling. Trends Pharmacol Sci 28(4):173–179

8. Luttrell LM (2008) Reviews in molecular biology and biotechnology: transmembrane signaling by G protein-coupled receptors. Mol Biotechnol 39(3):239–264

9. Magalhaes AC et al (2010) CRF receptor 1 regulates anxiety behavior via sensitization of

5-HT2 receptor signaling. Nat Neurosci 13 (5):622–629

10. Thompson MD, Cole DE, Jose PA (2008) Pharmacogenomics of G protein-coupled receptor signaling: insights from health and disease. Methods Mol Biol 448:77–107

11. Overington JP, Al-Lazikani B, Hopkins AL (2006) How many drug targets are there? Nat Rev Drug Discov 5(12):993–996

12. Jaferi A, Lane DA, Pickel VM (2009) Subcellular plasticity of the corticotropin-releasing factor receptor in dendrites of the mouse bed nucleus of the stria terminalis following chronic opiate exposure. Neuroscience 163 (1):143–154

13. Reyes BA et al (2006) Agonist-induced internalization of corticotropin-releasing factor receptors in noradrenergic neurons of the rat locus coeruleus. Eur J Neurosci 23 (11):2991–2998

14. Reyes BA, Valentino RJ, Van Bockstaele EJ (2008) Stress-induced intracellular trafficking of corticotropin-releasing factor receptors in rat locus coeruleus neurons. Endocrinology 149(1):122–130

15. Treweek JB et al (2009) Electron microscopic localization of corticotropin-releasing factor (CRF) and CRF receptor in rat and mouse central nucleus of the amygdala. J Comp Neurol 512(3):323–335

16. Williams TJ et al (2011) Ovarian hormones influence corticotropin releasing factor receptor colocalization with delta opioid receptors in CA1 pyramidal cell dendrites. Exp Neurol 230 (2):186–196

17. Leranth C, Pickel VM (1989) Electron microscopic preembedding double-labeling methods. In: Heimer L, Zaborszky L (eds) Neuroanatomical tracing methods 2, 1st edn. Plenum, New York, pp 129–172

18. Lujan R et al (1996) Perisynaptic location of metabotropic glutamate receptors mGluR1 and mGluR5 on dendrites and dendritic spines in the rat hippocampus. Eur J Neurosci 8 (7):1488–1500

19. Jin J et al (2010) Interaction of the mu-opioid receptor with GPR177 (Wntless) inhibits Wnt secretion: potential implications for opioid dependence. BMC Neurosci 11:33

20. Reyes AR et al (2010) Ultrastructural relationship between the mu opioid receptor and its interacting protein, GPR177, in striatal neurons. Brain Res 1358:71–80

21. Reyes BA et al (2012) Opiate agonist-induced re-distribution of Wntless, a mu-opioid receptor interacting protein, in rat striatal neurons. Exp Neurol 233(1):205–213

22. Yi H et al (2001) A novel procedure for pre-embedding double immunogold-silver labeling at the ultrastructural level. J Histochem Cytochem 49(3):279–284

23. Reyes BA et al (2011) Amygdalar peptidergic circuits regulating noradrenergic locus coeruleus neurons: linking limbic and arousal centers. Exp Neurol 230(1):96–105

24. Gainetdinov RR et al (2003) Dopaminergic supersensitivity in G protein-coupled receptor kinase 6-deficient mice. Neuron 38 (2):291–303

25. von Zastrow M (2003) Mechanisms regulating membrane trafficking of G protein-coupled receptors in the endocytic pathway. Life Sci 74 (2–3):217–224

26. von Zastrow M et al (2003) Regulated endocytosis of opioid receptors: cellular mechanisms and proposed roles in physiological adaptation to opiate drugs. Curr Opin Neurobiol 13 (3):348–353

27. Ferguson SS et al (1996) G-protein-coupled receptor regulation: role of G-protein-coupled receptor kinases and arrestins. Can J Physiol Pharmacol 74(10):1095–1110

28. Ferguson SS et al (1996) G-protein-coupled receptor kinases and arrestins: regulators of G-protein-coupled receptor sequestration. Biochem Soc Trans 24(4):953–959

29. Tsao P, von Zastrow M (2000) Downregulation of G protein-coupled receptors. Curr Opin Neurobiol 10(3):365–369

30. Gyombolai P et al (2013) Differential beta-arrestin2 requirements for constitutive and agonist-induced internalization of the CB1 cannabinoid receptor. Mol Cell Endocrinol 372(1–2):116–127

31. Hinkle PM, Gehret AU, Jones BW (2012) Desensitization, trafficking, and resensitization of the pituitary thyrotropin-releasing hormone receptor. Front Neurosci 6:180

32. Raposo G et al (1989) Internalization of beta-adrenergic receptor in A431 cells involves non-coated vesicles. Eur J Cell Biol 50(2):340–352

33. Tolbert LM, Lameh J (1996) Human muscarinic cholinergic receptor Hm1 internalizes via clathrin-coated vesicles. J Biol Chem 271 (29):17335–17342

34. Van Bockstaele EJ, Commons KG (2001) Internalization of mu-opioid receptors produced by etorphine in the rat locus coeruleus. Neuroscience 108(3):467–477

35. Castelli MP et al (1997) Chronic morphine and naltrexone fail to modify mu-opioid receptor mRNA levels in the rat brain. Brain Res Mol Brain Res 45(1):149–153

36. Pert CB, Snyder SH (1975) Identification of opiate receptor binding in intact animals. Life Sci 16(10):1623–1634

37. Petruzzi R et al (1997) The effects of repeated morphine exposure on mu opioid receptor number and affinity in C57BL/6J and DBA/2J mice. Life Sci 61(20):2057–2064

38. Arttamangkul S et al (2008) Differential activation and trafficking of micro-opioid receptors in brain slices. Mol Pharmacol 74(4):972–979

39. Johnson EA et al (2006) Agonist-selective mechanisms of mu-opioid receptor desensitization in human embryonic kidney 293 cells. Mol Pharmacol 70(2):676–685

40. Bailey CP et al (2003) Mu-opioid receptor desensitization in mature rat neurons: lack of interaction between DAMGO and morphine. J Neurosci 23(33):10515–10520

41. Sternini C et al (1996) Cellular localization of Pan-trk immunoreactivity and trkC mRNA in the enteric nervous system. J Comp Neurol 368(4):597–607

42. Emmerson PJ et al (1996) Characterization of opioid agonist efficacy in a C6 glioma cell line expressing the mu opioid receptor. J Pharmacol Exp Ther 278(3):1121–1127

43. Keith DE et al (1996) Morphine activates opioid receptors without causing their rapid internalization. J Biol Chem 271(32):19021–19024

44. Pak Y et al (1999) Agonist-induced G protein-dependent and -independent down-regulation of the mu opioid receptor. The receptor is a direct substrate for protein-tyrosine kinase. J Biol Chem 274(39):27610–27616

45. Selley DE et al (1997) Opioid receptor-coupled G-proteins in rat locus coeruleus membranes: decrease in activity after chronic morphine treatment. Brain Res 746(1–2):10–18

46. Keith DE et al (1998) mu-Opioid receptor internalization: opiate drugs have differential effects on a conserved endocytic mechanism in vitro and in the mammalian brain. Mol Pharmacol 53(3):377–384

47. Laporte SA et al (2000) The interaction of beta-arrestin with the AP-2 adaptor is required for the clustering of beta 2-adrenergic receptor into clathrin-coated pits. J Biol Chem 275(30):23120–23126

48. Laporte SA et al (1999) The beta2-adrenergic receptor/betaarrestin complex recruits the clathrin adaptor AP-2 during endocytosis. Proc Natl Acad Sci U S A 96(7):3712–3717

49. Ferguson G, Watterson KR, Palmer TM (2000) Subtype-specific kinetics of inhibitory adenosine receptor internalization are determined by sensitivity to phosphorylation by G protein-coupled receptor kinases. Mol Pharmacol 57(3):546–552

50. Ferguson G, Watterson KR, Palmer TM (2002) Subtype-specific regulation of receptor internalization and recycling by the carboxyl-terminal domains of the human A1 and rat A3 adenosine receptors: consequences for agonist-stimulated translocation of arrestin3. Biochemistry 41(50):14748–14761

51. Haberstock-Debic H et al (2005) Morphine promotes rapid, arrestin-dependent endocytosis of mu-opioid receptors in striatal neurons. J Neurosci 25(34):7847–7857

52. Haberstock-Debic H et al (2003) Morphine acutely regulates opioid receptor trafficking selectively in dendrites of nucleus accumbens neurons. J Neurosci 23(10):4324–4332

53. McPherson J et al (2010) mu-opioid receptors: correlation of agonist efficacy for signalling with ability to activate internalization. Mol Pharmacol 78(4):756–766

54. Melief EJ et al (2010) Ligand-directed c-Jun N-terminal kinase activation disrupts opioid receptor signaling. Proc Natl Acad Sci U S A 107(25):11608–11613

55. Whistler JL et al (1999) Functional dissociation of mu opioid receptor signaling and endocytosis: implications for the biology of opiate tolerance and addiction. Neuron 23(4):737–746

56. Jaremko KM et al (2014) Morphine-induced trafficking of a mu-opioid receptor interacting protein in rat locus coeruleus neurons. Prog Neuropsychopharmacol Biol Psychiatry 50:53–65

57. Bailey JE et al (2011) Preliminary evidence of anxiolytic effects of the CRF1 receptor antagonist R317573 in the 7.5% CO_2 proof-of-concept experimental model of human anxiety. J Psychopharmacol 25(9):1199–1206

58. Banziger C et al (2006) Wntless, a conserved membrane protein dedicated to the secretion of Wnt proteins from signaling cells. Cell 125(3):509–522

59. Saper CB, Sawchenko PE (2003) Magic peptides, magic antibodies: guidelines for appropriate controls for immunohistochemistry. J Comp Neurol 465(2):161–163

60. Jin J et al (2010) Expression of GPR177 (Wntless/Evi/Sprinter), a highly conserved Wnt-transport protein, in rat tissues, zebrafish embryos, and cultured human cells. Dev Dyn 239(9):2426–2434

61. Kravets JL et al (2015) Direct targeting of peptidergic amygdalar neurons by

noradrenergic afferents: linking stress-integrative circuitry. Brain Struct Funct 220 (1):541–558

62. Milner TA et al (2011) Degenerating processes identified by electron microscopic immunocytochemical methods. In: Manfredi G, Kawamata H (eds). Neurodegeneration: methods and protocols. Methods in molecular biology. Springer Science + Business Media, LLC

63. Peters A, Palay SL (1996) The morphology of synapses. J Neurocytol 25:687–700

64. Peters A, Palay SL, Webster HD (1991) The fine structure of the nervous system. Oxford University Press, New York

65. Chan J, Aoki C, Pickel VM (1990) Optimization of differential immunogold-silver and peroxidase labeling with maintenance of ultrastructure in brain sections before plastic embedding. J Neurosci Methods 33 (2–3):113–127

66. Scavone JL, Van Bockstaele EJ (2009) Mu-opioid receptor redistribution in the locus coeruleus upon precipitation of withdrawal in opiate-dependent rats. Anat Rec (Hoboken) 292(3):401–411

67. Reyes BA, Chavkin C, Van Bockstaele EJ (2010) Agonist-induced internalization of kappa-opioid receptors in noradrenergic neurons of the rat locus coeruleus. J Chem Neuroanat 40(4):301–309

68. Wang Y et al (2009) Effects of acute agonist treatment on subcellular distribution of kappa opioid receptor in rat spinal cord. J Neurosci Res 87(7):1695–1702

69. Fritschy JM (2008) Is my antibody-staining specific? How to deal with pitfalls of immunohistochemistry. Eur J Neurosci 28(12): 2365–2370

70. Rhodes KJ, Trimmer JS (2006) Antibodies as valuable neuroscience research tools versus reagents of mass distraction. J Neurosci 26 (31):8017–8020

71. Saper CB (2005) An open letter to our readers on the use of antibodies. J Comp Neurol 493(4):477–478

Neuromethods (2016) 115: 167–180
DOI 10.1007/7657_2015_76
© Springer Science+Business Media New York 2016
Published online: 23 February 2016

Identification of Select Autonomic and Motor Neurons in the Rat Spinal Cord Using Retrograde Labeling Techniques and Post-embedding Immunogold Detection of Fluorogold in the Electron Microscope

Leif A. Havton, Huiyi H. Chang, and Lisa Wulund

Abstract

It is of interest to combine intracellular neuronal labeling techniques with immunohistochemistry for ultrastructural studies, but experimental options for investigators have been limited. Recently, protocols have been developed to allow for electron microscopic detection of retrogradely labeled spinal cord neurons in combination with post-embedding immunogold studies. Retrogradely labeled preganglionic sympathetic and parasympathetic neurons are identified in the rat spinal cord after injection of fluorogold (FG) into the major pelvic ganglia. Similarly, injection of FG into the rat external urethral sphincter muscle retrogradely labels motoneurons in the dorsolateral (DL) nucleus of the L6 spinal cord segment. Tissues are next preserved by intravascular perfusion with a fixative solution containing 2 % paraformaldehyde and 1–2 % glutaraldehyde. Vibratome sectioning of the spinal cord tissue is followed by cryo-protection, snap freezing, and dehydration in methanol containing 1–2 % uranyl acetate using a freeze substitution protocol. Finally, the sections are embedded in the Lowicryl HM20 resin at low temperatures. Post-embedding immunogold detection of FG localizes the tracer in the lysosomes of labeled neurons. The protocol is versatile and can be combined with immunogold detection of, e.g., neurotransmitters and membrane transporters and therefore may have broad applicability to ultrastructural studies on the central nervous system.

Keywords: Ultrastructure, Major pelvic ganglion, External urethral sphincter, Anatomical tracer

1 Introduction

Significant progress has been demonstrated in electron microscopy over the past three decades with regard to the development and refinement of sensitive post-embedding immuno-gold methods. These protocols have allowed for the ultrastructural detection of amino acids, peptides, and proteins. Early studies provided methods for the detection of amino acids associated with synaptic transmission, including glutamate, L-aspartate, gamma-aminobutyric acid (GABA), and glycine [1–4]. Next, the introduction of the freeze substitution technique with use of Lowicryl HM20 for slow embedding of fixed brain tissues at low temperatures resulted in markedly improved antigen preservation [5]. In addition to the detection of amino acid transmitters, the freeze substitution approach and tissue embedding using Lowicryl HM20 also allow for post-embedding immunogold detection of neuropeptides and proteins, as has been shown in ultrastructural studies of a variety of neurotransmitters and their transporters as well as proteins associated with postsynaptic receptors in the brain and spinal cord [6–9].

In spite of the above significant advancements in the area of tissue preparation and processing for electron microscopy, a remaining methodological challenge has been to identify or develop a neuronal labeling protocol, which may be combined with post-embedding immunogold investigations. Many of the most commonly used anterogradely and retrogradely transported anatomical tracers include pre-embedding steps, which are not compatible with post-embedding immunogold studies. For instance, horseradish peroxidase (HRP) and HRP conjugated to wheat-germ agglutinin (WGA-HRP) or the B fragment of the cholera toxin (B-HRP) commonly use diaminobenzidine (DAB) or tetramethyl benzidine (TMB) as chromogen as well as osmium tetroxide to enhance contrast and detection of the reaction product [10–15]. Unfortunately, these pre-embedding histochemical steps markedly compromise antigen detection.

Here, we present a detailed description of a method that allows for post-embedding immunogold detection of retrogradely transported fluorogold (FG) in select motor and autonomic neurons in the rat spinal cord [16–19]. The method is based on the injection of FG into the major pelvic ganglion or external urethral sphincter muscle in rats (Figs. 1 and 2) and subsequent ultrastructural studies to identify retrogradely labeled autonomic and motor neurons after immunogold labeling of FG (Figs. 3 and 4).

2   Materials

The following list includes information regarding animals, surgical supplies, anatomical tracer, freeze substitution and embedding equipment, accessories, and supplies for immunogold studies.

1. Female Sprague-Dawley rats (180–220 g body weight; Charles River Laboratories, Raleigh, NC).

2. Isoflurane.

3. Betadine® surgical scrub solution (povidone-iodine, 7.5 %).

4. Sterile wipes saturated with 70 % isopropyl alcohol.

5. Sterile cotton swaps.

6. Standard scalpel blade holder (#3).

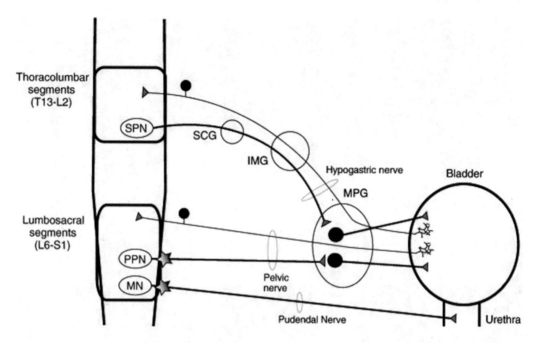

Fig. 1 Schematic representation of the innervation of the lower urinary tract in rats. Note that the major pelvic ganglion (MPG) is innervated by sympathetic preganglionic neurons (SPN) in the T13–L2 segments and parasympathetic preganglionic neurons (PPN) in the L6–S1 segment. Motoneurons (MN) in the L6–S1 segments innervate skeletal muscle fibers of the external urethral sphincter (reprinted from ref. [18] with permission from Elsevier)

7. Scalpel blade #10.

8. Retractor.

9. 10 µL Hamilton glass syringe with 32-gauge needle (Hamilton Company, Reno, NV).

10. 1.0 µl of a 5 % FG solution in sterile water (Fluorochrome, Denver, CO).

11. Surgical sutures.

12. Syringes.

13. Plastic pipettes.

14. Leica EM AFS2 (Leica Microsystems, Vienna, Austria).

15. Leica EM CPC (Leica Microsystems, Vienna, Austria).

16. Cryotransfer container with cover (Leica code 16701889).

17. Chamber for gelatin capsules (Leica code 16702790).

18. Chamber for Leica capsules (Leica code 16702789) with volume displacement body (Leica code 16702788).

19. Spider cover with eight positions (Leica code 16702743).

20. Stem holder for spider cover (Leica code 16702744).

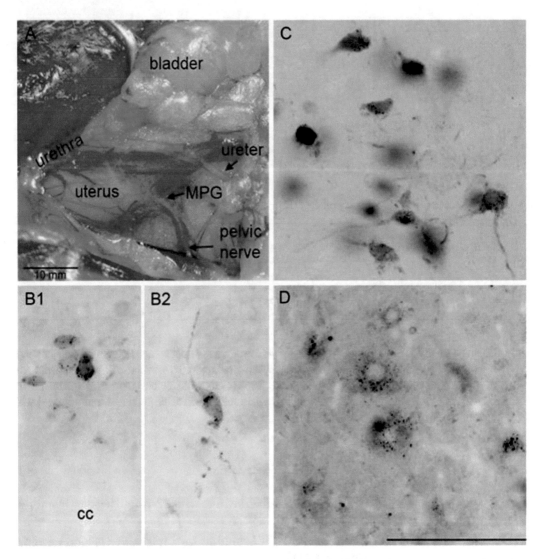

Fig. 2 Appearance of the major pelvic ganglion (MPG) in a female rat after the ganglion has been injected with a 5 % fluorogold solution (**a**). Note close relationship between the MPG and adjacent urethra, ureter, and pelvic nerve. Retrogradely labeled sympathetic preganglionic neurons are detected in the dorsal commissural nucleus (**b1**) and intermediolateral nucleus (**b2**) of the L2 segment, and retrogradely labeled preganglionic parasympathetic neurons (**c**) are present in the intermediolateral nucleus of the S1 segment. In addition, retrogradely labeled motoneurons (**d**) can be encountered in the dorsolateral nucleus of the L6 segment after a direct tracer injection into the external urethral sphincter muscle (reprinted from ref. [17] with permission from Elsevier)

21. Plastic capsules, D5XH15 mm, 200 pcs (Leica code 16702738).

22. Gelatin capsules size 1, 1,000 pcs (Leica code 16702745).

23. Special forceps with insulation coating (Leica code 16701955).

24. Universal containers (Leica code 16702735).

25. Cryotool with M4 thread (Leica code 16701958).

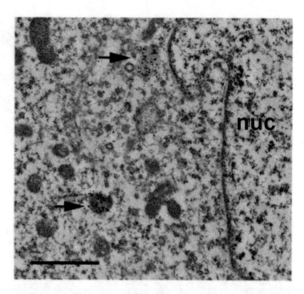

Fig. 3 Accumulation of FG in lysosome-like organelles in retrogradely labeled sympathetic preganglionic neuron (SPN) is detected by post-embedding immunogold studies and indicated by *arrows* in the cytoplasm of the SPN in near proximity of its nucleus (nuc). Scale bar = 1 μm (reprinted from ref. [19] with permission from Elsevier)

26. Special Allen key with handle 3 mm (Leica code 16870070).
27. Lowicryl HM20 (Electron Microscopy Sciences, Fort Washington, PA).
28. Powertome X Ultramicrotome (RMC Products, Boeckeler Instruments, Tucson, AZ).
29. Formvar.
30. Nickel grids for collection of ultrathin sections.
31. Uranyl acetate.
32. Methanol.
33. Ethanol.
34. Sodium borohydride.
35. Glycine.
36. Tris-buffered saline.
37. Triton X-100.
38. Fluorogold (Fluorochrome, Denver, CO).
39. Polyclonal antibody to fluorogold raised in rabbit (Fluorochrome).
40. Secondary goat anti-rabbit antibody conjugated to 15 nm colloidal gold particles.
41. Lead citrate.
42. Humid chambers.
43. Transmission electron microscope.

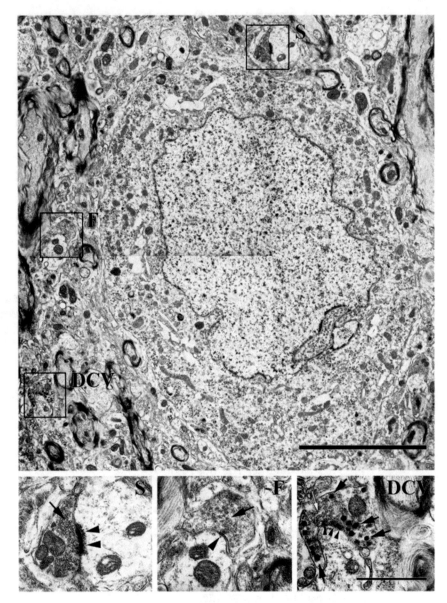

Fig. 4 Retrogradely labeled sympathetic preganglionic neuron after FG injection into the major pelvic ganglion. The FG is detected by post-embedding immunogold techniques. Note that the embedding procedures using freeze substitution techniques and immunogold labeling allow for ultrastructural preservation of the neurons, neuropil, and synaptic contacts for classification of presynaptic boutons containing spherical vesicles (S), flat vesicles (F), or dense core vesicles (DCV). Scale bar = 5 μm for composite image of whole SPN. Scale bar = 1 μm for bouton images (reprinted from ref. [19] with permission from Elsevier)

3 Methods

3.1 Surgical Procedures and Tracer Injections

1. Adult female rats are anesthetized for surgery using gas anesthesia (2–2.5 % isoflurane with oxygen) (*see* **Note 1**). The lower abdomen is shaved and prepped for surgery using three alternating applications of Betadine® surgical scrub and sterile wipes saturated with 70 % isopropyl alcohol.

2. A midline incision is made over the lower abdomen using a #3 scalpel equipped with a #10 scalpel blade. A retractor is placed in the surgical site to retract the abdominal wall muscles and expose the pelvic organs.

3. The bladder, urethra, and left ureter are dissected free and identified by using cotton swabs for blunt dissections (*see* **Note 2**). The left pelvic nerve is identified using a combination of anatomical landmarks, including the ureter and urethra. The MPG is then identified at the distal portion of the pelvic nerve (*see* **Note 3**).

4. A 32-gauge needle is attached to a 10 μl Hamilton syringe. The syringe is next filled with 1.0 μl of a 5 % solution of FG (Fluorochrome, Denver, CO) in sterile water, and the solution is injected into the MPG. Following the tracer injection, the needle is held in place for 1 min before being withdrawn. The injection site was next rinsed with saline (*see* **Notes 4** and **5**).

5. For retrograde labeling of motoneurons innervating the external urethral sphincter muscle, a total of 1.5 μl of the 5 % FG solution is injected into three sites of the muscle (*see* **Notes 6**).

6. The animals are allowed to recover from the surgical procedure and postoperative care is according to institutional guidelines and includes postoperative pain control, prophylactic antibiotic treatment, and monitoring for urinary retention.

3.2 Tissue Preservation, Cryo-Protection, and Embedding in Lowicryl HM20 Using a Freeze Substitution Protocol

1. At 5 days after the tracer injection, an over 1-week-long process of tissue preservation and embedding of spinal cord sections in a plastic resin takes place to allow for subsequent post-embedding immunogold labeling and ultrastructural studies (*see* **Note 7**). The beginning of this process takes place on day 0 and consists of the intravascular perfusion of the rats using an initial rinse of 50 ml of phosphate-buffered saline (PBS, pH 7.4) followed by 250 ml of a fixative solution containing 2 % paraformaldehyde + 1–2 % glutaraldehyde in phosphate buffer at room temperature (*see* **Note 8**). The spine is next removed and the thoracolumbar or lumbosacral portion of the spinal cord is dissected out. An oscillating tissue slicer (Electron Microscopy Sciences, Fort Washington, PA) is used for cutting transverse spinal cord sections at a thickness of 250 μm in ice-cold PBS (*see* **Note 9**). Next, trim each section to include the

Table 1
Leica EM AFS2 settings for freeze substitution

1.	−90 °C for 35 h.
2.	Raise 5 °C per hour for 9 h.
3.	−45 °C for 51 h.
4.	−45 °C for 27 h.
5.	Raise 5 °C per hour for 9 h.
6.	+/−0 °C for 40 h (24 h may be added here).

area of interest (*see* **Note 10**). The sections subsequently undergo freeze protection in a graded series of glycerol using 10, 20, and 30 % glycerol in PBS for 30 min each, followed by fresh 30 % glycerol in PBS overnight. Also on day 0, mix 1 % uranyl acetate in 100 % methanol and set aside for the following day.

2. At the beginning of day 1, fill the Leica EM CPC cryopreparation chamber and Leica EM AFS2 dewars with liquid nitrogen (*see* **Notes 11** and **12**). Next, program the Leica EM AFS2 according to the instructions in Table 1.

    When the programming is completed, initiate cooling to −90 °C and press PAUSE when the target temperature has been reached. Next, place the clean and dry cryo-chamber in the Leica EM CPC using the cryotool and push the chamber all the way down. Switch ON the Leica EM CPC. Set the temperature to −180 °C and start the liquid nitrogen pump.

3. Place the chamber for Leica capsules with the volume displacement body and eight plastic capsules inside the Leica EM AFS2 to chill. Place a universal chamber with 1 % uranyl acetate inside the Leica EM AFS2 and fill the capsules to the volume displacement body with uranyl acetate using a disposable plastic pipette.

    When the Leica CPC is chilled to about −180 °C, connect the propane cylinder to the propane hose and begin filling the propane chamber. Fill the propane chamber up to the groove by opening the pressure reduction valve slowly. When it is filled, press the Transfer Key (TF) and the CFP control panel to flood the CPC with liquid nitrogen to maintain the low temperature and carefully lift the propane chamber until it "clicks" in place.

4. For plunge freezing of tissue sections, mount a single section onto the pin on the injector rod. Next, place the rod in the ejector and "fire." Lift the pin with the section carefully, eject it into the cryotransfer container, and repeat the plunge freeze procedure for additional sections until the plunge freezing

series is completed. Remove each section from its pin by using a chilled spatula and set of tweezers to drop each section into a capsule containing a solution of 1 % uranyl acetate in methanol.

5. Close the lid of the cold chamber of the Leica EM AFS2 and press START to activate the program (*see* **Note 13**).

6. Remove the propane chamber from the Leica EM CPC and place it in the base cylinder of the propane torch to begin procedure for propane elimination (*see* **Note 14**).

7. Begin defrosting the Leica EM CPC by setting the temperature to −150 °C. When initial defrosting is completed, press the Heat (H) key of the Leica EM CPC to heat it up to room temperature.

8. Dispose of used and residual solutions containing uranyl acetate in proper waste containment according to institutional guidelines and soak exposed containers in ethanol. This marks the end of day 1.

9. There are no activities on day 2, as the sections are undergoing dehydration by the Leica EM AFS2.

10. At the beginning of day 3, fill a universal chamber with 100 % methanol and let it chill in the Leica EM AFS2.

11. Mix Lowicryl HM20 in fume hood using the recipe in Table 2 (*see* **Note 15**).

12. Remove the uranyl acetate-containing methanol solution from the capsules in the chamber for plastic capsules. Next, rinse the sections in capsules with methanol for 10 min at least three times.

13. Remove the last rinse of methanol from the plastic capsules and proceed with the infiltration steps according to the schedule in Table 3.

Table 2
Lowicryl HM20 recipe

Cross-linker D	2.98 g
Monomer E	17.02 g
Initiator C	0.10 g

Table 3
Lowicryl HM20:methanol ratios for tissue infiltration

1:1	1–2 h
2:1	1–2 h
Pure Lowicryl HM20	1–2 h
Pure Lowicryl HM20	Overnight

14. At the beginning of day 4, fill the chamber for gelatin capsules with the long portion of gelatin capsules and put into the Leica EM AFS2 to chill. The chamber holds a total of eight gelatin capsules. Next, fill the gelatin capsules with fresh, chilled Low-icryl HM20. Transfer the sections from the Leica capsules to the gelatin capsules and position each section at the bottom of the gelatin capsule. Drop in labels for each capsule with label positioned against the capsule wall (*see* **Note 16**).

15. Place stem holder for spider cover into universal container filled with chilled ethanol. Next, press spider cover onto the opening of the gelatin capsules in the chamber for gelatin capsule using specialized tool. Transfer the spider cover with attached gelatin capsule from the chamber for gelatin capsules to the ethanol-containing universal container and place the spider cover onto the stem holder for spider cover (*see* **Note 17**).

16. Attach the ultraviolet (UV) light to the Leica EM AFS2 and turn on the light to polymerize the plastic for 30 h at −45 °C followed by an increase in the temperature by 5 °C per hour for 9 h. Next, allow continued polymerization in the presence of UV light for 40 h.

17. There are no activities on day 5 or 6, as the Leica EM AFS automatically performs the polymerization process at low temperatures and in the presence of UV light.

18. On day 7 or 8, shut down the Leica EM AFS2 and remove the polymerized capsules.

3.3 Ultrathin Sectioning, Post-embedding Immunogold Labeling, and Ultrastructural Studies

1. Choose embedded tissue blocks containing the area of interest. If needed, trim the bottom portion of the tissue block containing the tissue to a smaller size, so that subsequent sections of the primary area of interest can comfortably can fit within the size of a one-hole nickel grid. Use an ultramicrotome to perform ultrathin sectioning of the area of interest and collect the sections on formvar-coated nickel grids (*see* **Note 18**).

2. For immunogold detection of FG, incubate the section-containing grids in drops containing the reagents and with the drops placed inside a humid chamber.

3. First, place the grids into drops containing 0.1 % sodium borohydride + 50 mM glycine in Tris-buffered saline (pH: 7.4) with 0.1 % Triton X-100 (TBST) for 25 min (*see* **Note 19**).

4. Rinse the grids three times in TBST and next incubate the sections in TBST containing normal goat serum (NGS, 1:20) for 10 min.

5. Next, incubate the grids with a polyclonal anti-rabbit primary FG antibody (1:20, Fluorochrome) in Tris buffer with 0.1 % Triton X-100 at 4 °C for 48 h.

6. Rinse the grids again in TBST and incubate the sections in TBST with NGS (1:20) for 10 min.

7. Incubate the grids at room temperature for 4 h with a secondary goat anti-rabbit antibody, which has been conjugated to 15 nm colloidal gold particles (1:10 GE Healthcare, Buckinghamshire, UK).

8. Rinse the grids in ddH$_2$O. Let the grids air-dry and next contrast the sections with 8 % uranyl acetate and lead citrate.

9. Finally, examine the immuno-stained and contrasted ultrathin sections in a transmission electron microscope for detection of immunogold labeling of FG in preganglionic autonomic and motor neurons (*see* **Note 20**).

---

4   Notes

1. A surgical plane of anesthesia can be confirmed by applying a toe pinch with a pair of forceps to make sure that no limb withdrawal takes place in response to a noxious stimulus.

2. The use of cotton swabs allows for gentle removal of connective and fat tissues, which may obscure the urethra and nervous structures.

3. MPG is located near the area where the ureter enters the bladder, and at the lateral side of urethra.

4. Although a 33-gauge needle is in some ways nearly perfect with its smaller tip for MPG injection, the needle is too soft to insert into MPG. Instead, use a 32-gauge needle, and the bevel should be totally inserted into the MPG to prevent leakage.

5. A bump is formed by injecting the solution to the MPG. After withdrawing the needle, the bump stays. It is a sign that the MPG injection was successful.

6. The urethra is identified below the bladder and the bladder neck. The external urethral sphincter muscle can be visualized as it surrounds the urethra. The middle portion of the urethra is ideal for the tracer injection into the external urethral sphincter muscle because of the thicker muscle layer here.

7. Many tissue preservation protocols are harsh on cellular proteins and epitopes detectable by immunohistochemistry. The present freeze substitution protocol allows for preservation of protein integrity and plastic embedding of tissues using slow tissue infiltration of the plastic resin at low temperatures in the absence of osmium exposure.

8. FG can be detected using post-embedding immunogold studies using a relatively wide range of aldehyde fixatives. For immunogold studies using other primary antibodies, exploratory studies are suggested to determine the degree of aldehyde fixation that is compatible with epitope detection.

9. For light microscopy and detection of the fluorescent marker (FG) in labeled neurons, thinner sections of 100 μm thickness are suggested. After injection of FG into the major pelvic ganglion, tissues are collected from the L1–L3 and L6–S1 spinal cord segments to include autonomic preganglionic neurons. After injection of FG into the external urethral sphincter, tissues are collected from the L6–S1 segments to include the dorsolateral nucleus, which primarily is located within the L6 segment.

10. The trimmed sections need to be small enough to fit at the bottom of the gelatin capsules used in later steps during the plastic infiltration steps.

11. The Leica EM CPC is a cryopreparation chamber that allows for plunge freezing of biological tissues. The Leica EM AFS2 performs freeze substitution and is used for progressive lowering of temperature techniques to allow for tissue embedding and polymerization of resins at low temperatures.

12. The following is a good resource for accessories available for the Leica EM CPC and Leica EM AFS2: http://www.leica-micro systems.com/products/em-sample-prep/biological-specimens/ low-temperature-techniques/tissue-processing/details/product/ leica-em-afs2/catalog/

13. The purpose of this step is to replace the water in the tissues with methanol in the presence of uranyl acetate. The Leica EM AFS2 allows this process to be performed slowly at low temperatures. The dehydration will initially take place at −90 °C for 35 h. Next the temperature is raised by 5 °C per hour for 9 h to −45 °C.

14. We recommend that each research laboratory contacts its institution's Office for Environmental Health and Safety to learn about specific and local regulations for the safe use and elimination of propane. The propane torch may be used to burn the remaining propane, but consultation with institutional guidelines and officials are again recommended to assure of adherence to institutional safety regulations.

15. Lowicryl HM20 is an odorous organic solvent used for plastic embedding of tissues. It should be mixed in a fume hood to minimize vapor exposures. Lowicryl HM20 is sensitive to oxygen, so avoid creating air bubbles when mixing the solution.

16. The printed labels are a convenient way of establishing a permanent identification of each section. Arial font at size 9 is suggested for the labels.

17. When placing the spider cover with gelatin capsules into the ethanol-containing chamber, make sure that the ethanol does not come into direct contact with the Lowicryl HM20. The ethanol will compromise polymerization of the plastic.

18. One-hole nickel grids coated with formvar will allow for unobstructed view of the tissue section in the electron microscope.

19. This initial incubation step aims at blocking free aldehyde residues in the tissues.

20. The immunogold labeling of FG is detected within lysosome-like structures in the cytoplasm of retrogradely labeled neurons.

Acknowledgements

Supported by grant funds from the Dr. Miriam and Sheldon G. Adelson Medical Research Foundation.

References

1. Ottersen OP (1987) Postembedding light and electron microscopic immunocytochemistry of amino acids: description of a new model system allowing identical conditions for specificity testing and tissue processing. Exp Brain Res 69:167–174

2. Örnung G, Shupliakov O, Lindå H, Ottersen OP, Storm-Mathisen J, Ulfhake B, Cullheim S (1996) Qualitative and quantitative analysis of glycine- and GABA-immunoreactive nerve terminals on motoneuron cell bodies in the cat spinal cord: a postembedding electron microscopic study. J Comp Neurol 365:413–426

3. Örnung G, Ottersen OP, Cullheim S, Ulfhake B (1998) Distribution of glutamate-, glycine- and GABA-immunoreactive nerve terminals on dendrites in the cat spinal motor nucleus. Exp Brain Res 118:517–532

4. Larsson M, Persson S, Ottersen OP, Broman J (2001) Quantitative analysis of immunogold labeling indicates low levels and non-vesicular localization of L-aspartate in rat primary afferent terminals. J Comp Neurol 430:147–159

5. Van Lookeren Campagne M, Oestreicher AB, van der Krift TP, Gispen WH, Verkleij AJ (1991) Freeze-substitution and Lowicryl HM20 embedding of fixed rat brain: suitability for immunogold ultrastructural localization of neural antigens. J Histochem Cytochem 39:1267–1279

6. Mahendrasingam S, Wallam CA, Hackney CM (2003) Two approaches to double postembedding immunogold labeling of freeze-substituted tissue embedded in low temperature Lowicryl HM20 resin. Brain Res Protoc 11:134–141

7. Persson S, Broman J (2004) Glutamate, but not aspartate, is enriched in trigeminothalamic tract terminals and associated with their synaptic vesicles in the rat nucleus submedius. Exp Brain Res 157:152–161

8. Larsson M, Broman J (2005) Different basal levels of CaMKII phosphorylated at $Thr^{286/287}$ at nociceptive and low-threshold primary afferent synapses. Eur J Neurosci 21:2445–2458

9. Larsson M, Broman J (2011) Synaptic plasticity and pain: role of ionotropic glutamate receptors. Neuroscientist 17:256–273

10. Mesulam MM (1978) Tetramethyl benzidine for horseradish peroxidase neurohistochemistry: a non-carcinogenic blue reaction product with superior sensitivity for visualizing neural afferents and efferents. J Histochem Cytochem 26:106–117

11. Olucha F, Martínez-García F, López-García C (1985) A new stabilizing agent for the tetramethyl benzidine (TMB) reaction product in

the histochemical detection of horseradish peroxidase (HRP). J Neurosci Methods 13:131–138

12. Ralston HJ (1990) Analysis of neuronal networks: a review of techniques for labeling axonal projections. J Electron Microsc Tech 15:322–331

13. Havton LA, Broman J (2005) Systemic administration of cholera toxin B subunit conjugated to horseradish peroxidase in the adult rat labels preganglionic autonomic neurons, motoneurons, and select primary afferents for light and electron microscopic studies. J Neurosci Methods 149:101–109

14. Ichiyama RM, Broman J, Edgerton VR, Havton LA (2006) Ultrastructural synaptic features differ between alpha- and gamma-motoneurons innervating the tibialis anterior muscle in the rat. J Comp Neurol 499:306–315

15. Ichiyama RM, Broman J, Roy RR, Zhong H, Edgerton VR, Havton LA (2011) Locomotor training maintains normal inhibitory influence on both alpha- and gamma-motoneurons after neonatal spinal cord transection. J Neurosci 31:26–33

16. Persson S, Havton LA (2009) Retrogradely transported fluorogold accumulates in lysosomes of neurons and is detectable ultrastructurally using post-embedding immuno-gold methods. J Neurosci Methods 184:42–47

17. Chang HY, Havton LA (2010) Anatomical tracer injections into the lower urinary tract may compromise cystometry and external urethral sphincter electromyography in female rats. Neuroscience 166:212–219

18. Wu L, Wu J, Chang HH, Havton LA (2012) Selective plasticity of primary afferent innervation to the dorsal horn and autonomic nuclei following lumbosacral ventral root avulsion and reimplantation in long term studies. Exp Neurol 233:758–766

19. Wu L, Chang HH, Havton LA (2012) The soma and proximal dendrites of sympathetic preganglionic neurons innervating the major pelvic ganglion in female rats receive predominantly inhibitory inputs. Neuroscience 217:32–45

Neuromethods (2016) 115: 181–203
DOI 10.1007/7657_2015_79
© Springer Science+Business Media New York 2015
Published online: 18 July 2015

Postembedding Immunohistochemistry for Inhibitory Neurotransmitters in Conjunction with Neuroanatomical Tracers

Miriam Barnerssoi and Paul J. May

Abstract

This chapter is aimed at providing the reader with detailed instructions for combining postembedding immunohistochemistry for inhibitory transmitters with methods for characterizing neuronal connections at the ultrastructural level. Thus, it includes protocols for both retrograde and anterograde neuronal tracing that have been modified to facilitate electron microscopic examination of neuronal connectivity. In addition, it includes protocols for doing immunohistochemistry with antibodies to either glutaraldehyde-fixed gamma amino butyric acid or glycine on ultrathin sections. By employing these procedures together one can identify both the input/output characteristics of a neuronal structure and identify whether these synaptic connections utilize one of these inhibitory transmitters.

Keywords Ultrastructure, GABA, Glycine, Synapse, Terminal, Neuronal circuits

1 Introduction

It has long been clear that the nervous system employs different subsystems for activation of neurons and for inhibiting this activity. In almost every system, the realization that activity is critically shaped by inhibitory processes has motivated examination of the inhibitory circuits crucial to the functioning of the connections being examined. The role of inhibition has been further emphasized by the realization that activity in circuits can be modulated both by local circuit inhibitory neurons and long axon pathways that also employ inhibition. In this context, it is not surprising that from early on those studying the ultrastructure of the central nervous system wished to be able to differentiate excitatory from inhibitory synapses. It was for this purpose that Gray [1] divided synaptic terminals into Type 1 and Type 2 terminals. The former, which were presumed to be excitatory, contained spherical vesicles and made clearly asymmetric synaptic contacts with prominent postsynaptic densities. The latter, which were presumed to be inhibitory, contained flattened vesicles and made symmetric synaptic contacts with modest postsynaptic densities. As useful as this classification scheme is, the fact remains that many terminals are not filled with patently flattened vesicles or homogeneously spherical vesicles. Instead, they contain variable numbers of pleomorphic vesicles making them difficult to peg as either inhibitory or excitatory. In addition, the degree of asymmetry in synaptic densities varies widely, and their classification is also affected by issues of plane of section [2, 3]. Thus, it has been very helpful that another approach to identifying inhibitory synapses, in the form of immunohistochemistry, has become available to resolve these issues at the ultrastructural level.

Immunohistochemical approaches originated from light microscopic studies that have used antibodies to elucidate the neurotransmitter specific architecture of the brain. Numerous investigators then took tissue in which the antibody was tagged with diaminobenzidine (DAB), and examined it at the electron microscopic level. If the goal is to simply identify the neurotransmitter in terminals in a sample, this approach is useful. However, most procedures for light microscopic immunohistochemistry utilize paraformaldehyde or formalin-fixed tissue. In addition, they employ detergents, usually 0.3 % Triton X-100, to partially solubilize cell membranes so that the large antibody complexes can gain access to the depths of the tissue. These two technical elements usually degrade the ultrastructure of the specimen, making it difficult to correlate the ultrastructure of labeled terminals and their postsynaptic targets with elements found in the normal ultrastructure of the region. The fixation problems can be ameliorated by including small amounts of glutaraldehyde in the perfusate (<0.2 %), or by postfixing in glutaraldehyde after the immunohistochemical steps. The detergent problem can be approached by using little or no Triton X-100, and then concentrating the ultrastructural examination on the outer few microns of the specimen.

A more direct approach to dealing with these shortcomings was taken by Somogyi and colleagues [4, 5]. They developed a post-embedding immunohistochemical technique to provide transmitter identification with superior ultrastructural preservation. The first key element was to develop antibodies to antigens that have been fixed with glutaraldehyde. Thus, the tissue can be fixed to provide superior ultrastructural detail, but the antibody still recognizes the neurotransmitter even after its conformation has been changed by the fixation process. The second key element was to develop an approach that allows a small portion of the embedding resin to be stripped away from the surface of the ultrathin section using hydrogen peroxide, so that the antigen is made available for attachment by the antibody. Somogyi and colleagues used DAB as an electron-dense tag for the antibody. While easy to visualize, this approach is more technically challenging to employ and it somewhat obscures the ultrastructure. More recently, many labs have begun to use a secondary antibody with a gold particle attached to it. This approach is somewhat simpler, and has the added advantage that it can be used in conjunction with neuroanatomical techniques that employ DAB as the electron-dense tag for neurons or terminals whose connectivity has been specified with tracers. Postembedding immunohistochemistry has been most widely used to identify inhibitory synapses and cells that contain either gamma amino butyric acid (GABA) or glycine as a transmitter. This is no doubt primarily due to the fact that antibodies to glutaraldehyde-fixed GABA and glycine are commercially available. However, it is also motivated by the chemical characteristics of these neurotransmitters. As small molecules, they are not well fixed by paraformaldehyde. Consequently, they tend to diffuse out of the tissue and thus present a difficult experimental target for conventional immunohistochemical techniques. Fixation with a perfusate containing between 1.5 and 2.0 % glutaraldehyde, in addition to 1.0 % paraformaldehyde, stabilizes these small molecules so that they can be found by the antibody. This approach has been used successfully in a wide variety of structures and systems [6–15].

Development of immunohistochemical techniques has proceeded in parallel with development of retrograde and anterograde neuronal tracer techniques. It has been particularly useful to combine these techniques in order to specify whether a particular cell population with a particular target utilizes a particular neurotransmitter [16–22]. Neuroanatomical tracers have been used at the electron microscopic level to allow the identification of both terminals via anterograde tracers and cells via retrograde tracers. More recently these techniques have been combined to specify monosynaptic input/output linkages between anterogradely identified terminals and retrogradely identified cell populations. Obviously it would be advantageous to further specify whether these linkages are inhibitory or excitatory. To do this, one can combine single or dual tracer neuroanatomical techniques and postembedding immunohistochemistry in order to allow terminals to be characterized at the electron microscopic level. We have had considerable success with this approach, and we offer here our procedures for its employment. Note, that this approach has also been used for glutamate. Since we have not used anti-glutamate antibodies, we will not discuss this further here, but the basic idea is the same.

Since individuals reading this chapter will come to it with different skill sets, we have organized the Materials and the Methods sections in a modular fashion, thus allowing the reader to pick and choose which elements they would like to attend to (*see* **Note 1**). An outline of these modules is provided here:

1. Perfusion and sectioning

2. Retrograde tracing

3. Anterograde tracing

4. Sampling, processing, and EM ultrathin cutting

5. Postembedding immunohistochemistry

6. Heavy metal staining

7. Analysis

2   Materials

2.1   Perfusion and Sectioning

1. Perfusion rinse: 0.1 M, pH 7.2 phosphate-buffered saline (PBS). Stock buffer concentrate, 0.4 M, pH 7.2 phosphate buffer (PB), contains 10.6 g of monobasic, monohydrate sodium phosphate [$NaH_2PO_4 \cdot H_2O$], and 56.0 g of dibasic potassium phosphate [K_2HPO_4] per 1.0 l of deionized water (dH_2O). Store refrigerated. This stock is diluted 1:3 with dH_2O to make 0.1 M working solution PB. PBS for the perfusion rinse is made by adding 0.85 g of NaCl per 100 ml of the working solution PB.

2. Mixed aldehyde fixative: 1.0 % paraformaldehyde and 1.5–2.0 % glutaraldehyde in 0.1 M, pH 7.2 PB (*see* **Note 2**). For each 1,000 ml of a 1.0 %/1.5 % mixture of fixative solution needed, place 10 g of paraformaldehyde [$(CH_2O)_n$] (Fisher T-353) into 100 ml of dH_2O in a flask. Then heat to 60 °C while stirring under a hood. Next, add 1.0 N NaOH, dropwise, until the solution clears (*see* **Note 3**). Cool below 30 °C and filter. For 1,000 ml of fixative, add 500 ml of dH_2O and 250 ml of 0.4 M, pH 7.2 PB to solution, and store at 4 °C. Before perfusion add 60 ml of 25 % glutaraldehyde [$OHC(CH_2)_3CHO$] (Fisher, 02957) and then 90 ml dH_2O measured in the same graduated cylinder for the 1.5 % glutaraldehyde fixative (*see* **Note 4**).

2.2   Retrograde Tracing

1. Retrograde tracers: 1.0–2.0 % Wheatgerm Agglutinin Conjugated to Horseradish Peroxidase (Sigma-Aldrich, L3892) (WGA-HRP) or 1.0–2.0 % Choleratoxin B-subunit conjugated to horseradish peroxidase (Sigma-Aldrich, C3741) (ChTB-HRP). Dissolved in dH_2O for both pressure and iontophoretic injections (*see* **Note 5**). Dissolve 0.005 g of tracer into 25 μl of dH_2O to make a 2.0 % solution.

2. Stock 0.4 M, pH 6.0 PB. This buffer contains 48.45 g of monobasic, monohydrate sodium phosphate [$NaH_2PO_4 \cdot H_2O$], and 8.56 g of dibasic potassium phosphate [K_2HPO_4] in 1.0 l of dH_2O. Store refrigerated. This is diluted 1:3 with dH_2O to make 0.1 M working solution (*see* **Note 6**).

3. Molybdate-TMB reaction solution A: This solution contains 1.95 g ammonium molybdate 4-hydrate, crystal [$NH_4Mo_7O_{24} \cdot 4H_2O$] in 780.0 ml of 0.1 M, pH 6.0 PB.

4. Molybdate-TMB reaction solution B: This solution contains 0.04 g of tetramethylbenzidine free base (TMB) $[C_6H_2(CH_3)_2 \cdot 4NH_2]_2$ (Sigma-Aldrich, T2885) in 20 ml of 100 % ethanol (*see* **Note 7**). Dissolve in covered beaker by stirring. Do NOT heat. This takes some time to go into solution. *N.B. TMB may be carcinogenic. All steps employing this chemical should be done using gloves and with the solutions sealed, for example, by placing plastic wrap over the reaction tray. Store used solution for proper disposal.*

5. 0.3 % Hydrogen peroxide: Dilute 0.3 ml of 30 % stock hydrogen peroxide $[H_2O_2]$ (Fisher, H325-100) in 45.5 ml of dH_2O. *Use gloves; do NOT mouth pipette.*

6. Stabilizer solution: 5.0 % Ammonium molybdate. Dissolve 30.0 g of ammonium molybdate 4 hydrate $[NH_4Mo_7O_{24} \cdot 4H_2O]$ into 600 ml of 0.1 M, pH 6.0 PB.

7. 0.1 M, pH 7.2 PB. See above.

8. Diaminobenzidine (DAB) protection solution: Dissolve 0.06 g of 3,3′-diaminobenzidine tetrahydrochloride $[C_{12}H_{14}N_4 \cdot 4HCl]$ (Sigma-Aldrich, D5637) (*see* **Note 8**) in 450 ml of dH_2O. Then mix with 150 ml of 0.1 M, pH 7.2 PB. *N.B. DAB is known to be carcinogenic. All steps employing this chemical should be done using gloves in a hood. Store used solution for proper disposal.*

2.3 Anterograde Tracing

1. Anterograde tracer: 10.0 % Biotinylated Dextran Amine, 10,000 MW (Invitrogen-Molecular Probes, D1956) (BDA). Dissolve BDA in dH_2O for both pressure and iontophoretic injections (*see* **Note 5**). Dissolve 0.005 g of tracer into 50 µl of dH_2O.

2. 0.1 M, pH 7.2 PB. See above.

3. Buffered 0.05 % Triton X-100: Stock solution of 0.3 % Triton X-100 is made by adding 0.3 ml of Triton X-100 [t-octylphenoxypolyethoxyethanol] (Sigma-Aldrich, A-100) into 99.7 ml of 0.1 M, pH 7.2 PB with a micropipette (*see* **Note 9**). The working concentration of 0.05 % is produced by diluting 15 ml of 0.3 % stock per 90 ml of 0.1 M, pH 7.2 PB.

4. 1:500 Avidin D-conjugated horseradish peroxidase solution (avidin-HRP): This is made by diluting 40 µl of avidin-HRP (Vector Labs, A2004) (*see* **Note 10**) into 20 ml of 0.05 % buffered Triton X-100.

5. Nickel cobalt diaminobenzidine solution: Dissolve 0.06 g of 3,3′-diaminobenzidine tetrahydrochloride $[C_{12}H_{14}N_4 \cdot 4HCl]$ (Sigma-Aldrich, D5637) (*see* **Note 8**) in 450 ml of dH_2O. Then, mix with 150 ml of 0.1 M, pH 7.2 PB. Immediately before using, stir in 6.0 ml of 1.0 % nickel ammonium sulfate

and 2.0–6.0 ml of 1.0 % cobalt chloride. These two 1.0 % stock solutions are made by dissolving 1 g of nickel ammonium sulfate [$NiSO_4(NH_4)_2SO_4 \cdot 6H_2O$] into 100 ml of dH_2O, and 1 g of cobalt chloride [$CoCl_3$] into 100 ml of dH_2O. Both stock solutions are stored at 4 °C. *N.B. DAB is known to be carcinogenic. All steps employing this chemical should be done using gloves in a hood. Store used solution for proper disposal.*

2.4 Sampling, Processing, and EM Ultrathin Cutting

1. 0.2 M, pH 7.0 EM Phosphate buffer (EM PB): For 100 ml add 1.229 g of monobasic, dihydrate sodium phosphate [$NaH_2PO_4 \cdot 2H_2O$] (Fisher Scientific, S-381), and 1.739 g of dibasic sodium phosphate [Na_2HPO_4] (Fisher Scientific, S-374) to 97 ml of dH_2O.

2. 1.0 % Osmium tetroxide in 0.1, pH 7.0 M PB: Stock 2.0 % osmium tetroxide solution is made up by dissolving 1 g of osmium tetroxide [OsO_4] (EM Sciences, 19100) in 50 ml of dH_2O. The working solution is produced by mixing 1 part of the 2.0 % solution with 1 part 0.2 M, pH 7.0 EM phosphate buffer. (Filter before use.) *Osmium tetroxide is extremely reactive. It should only be used in the hood while wearing gloves and other protective gear. Store used solution and vials in a sealed glass container for proper disposal.*

3. Durcupan ACM Epoxy (EM Sciences, 14040)

2.5 Postembedding Immunohisto-chemistry

The TBS, TBS + BSA, and TBS + BSA + Tween 20 described in solutions 2–4 below should be made up fresh every time tissue is processed for postembedding. The staining process takes 2 days. Keep solutions refrigerated between days 1 and 2. The glutaraldehyde solution described in **step 7** can be made up in bulk and stored (*see* **Note 14**).

1. 0.1 M, pH 7.0 EM Phosphate buffer: Dilute 0.2 M EMPB (see above) 1:1 with dH_2O.

2. 0.1 M, pH 7.4 Tris-buffered saline (TBS). Add 0.76 g of Trizma (Sigma, T7693) and 0.85 g of NaCl to 100 ml of dH_2O in order to produce 100 ml of TBS.

3. TBS + bovine serum albumin (BSA): Take 50 ml of freshly made TBS and add 0.5 g of BSA (Sigma, A7906). Do not shake, but sonicate to put into solution.

4. TBS + BSA + Tween 20 (polyoxyethylenesorbitanmonolaurate): Take 40 ml of freshly made TBS + BSA and add 0.2 µl of Tween 20 (Sigma, P-137). Do not shake. Instead, sonicate to put into solution.

5. Primary antibody. For anti-GABA, use antibody to glutaraldehyde-fixed GABA raised in rabbits (Sigma, A2052). Dilute antibody 1: 250 in TBS + BSA + Tween 20 by adding 10 µl of antibody to 2.5 ml of TBS + BSA + Tween 20 (*see* **Note 11**).

For anti-glycine, use antibody to glutaraldehyde-fixed glycine raised in rabbits (Millipore, AB5020). Dilute antibody, 1:10, in TBS+BSA+Tween 20 by adding 100 μl of antibody to 1.0 ml of TBS+BSA+Tween 20 (*see* **Note 12**).

6. Secondary antibody: Goat anti-rabbit IgG conjugated to 15 nm gold particles (EM Sciences, 25113 (Aurion immunogold reagents)) in TBS+BSA+Tween 20. Dilute antibody 1:75 by placing 20 μl of it into 1.5 ml of TBS+BSA+Tween 20 (*see* **Note 13**).

7. 2.0 % Glutaraldehyde in 0.1 M, pH 7.0 EM PB: For 25.0 ml, use 12.5 ml of previously made PB. Add 10.5 ml of dH$_2$O and 2.0 ml of 25 % glutaraldehyde (Fisher Scientific, 02957-1). We use conventional glutaraldehyde instead of glutaraldehyde exclusively produced for EM, as we could not detect a significant difference in ultrastructural preservation in the past.

2.6 Heavy Metal Staining

1. Uranyl acetate solution: Make a 2.0 % solution by placing 0.2 g of uranyl acetate [(UO$_2$(Ch$_3$COO)$_2$)] (Ted Pella, 19481) in 10 ml of distilled (dH$_2$O) water. Filter before use.

2. Lead citrate solution: Add 0.40 g of lead citrate 3-hydrate [Pb$_3$(C$_6$H$_5$O$_7$)$_2$·3H$_2$O] (Electron Microscopic Sciences 512-26-5), 0.30 g of lead nitrate [Pb(NO$_3$)$_2$] (Mallinckrodt, 5744), 0.30 g of lead acetate [Pb(C$_2$H$_3$O$_2$)$_2$·3H$_2$O] (Mallinckrodt, 5688), and 2.00 g of sodium citrate [Na$_3$C$_6$H$_5$O$_7$·2H$_2$O] (Mallinckrodt, 0754) to 82 ml of boiled, cooled dH$_2$O. Reagents are sonicated to place them into solution. The solution will remain cloudy until the sodium citrate is added. Add 18 ml of 1 N sodium hydroxide [NaOH]. Filter before use.

3 Methods

3.1 Perfusion and Sectioning

1. Animals are deeply anesthetized with sodium pentobarbital (50 mg/kg) only after the perfusion setup is entirely in place with all solutions loaded, and the surgeon has on protective eyewear, clothing (we prefer a plastic apron), mask, and gloves.

2. The perfusion rinse, 0.1 M, pH 7.2 PBS, is pumped through the vasculature by use of a peristaltic pump (Master Flex LS, Cole Palmer). To accomplish this, a blunt 13 G, 2¾ inch, stainless steel LNR needle (BD) attached to Tygon tubing with a Luer-Lock hub is advanced through the wall of the left ventricle, and threaded through the aortic valve until it lies in the base of the aorta. The needle is stabilized by clamping the ventricle around it with an Allis hemostatic forceps. The right atrium is opened with a fine scissors. The descending aorta is

clamped with a hemostat at the level just above the diaphragm, if one is studying the brain and not spinal cord.

3. Once the fluid leaving the atrium has cleared of blood, the fixative, 1.0 % paraformaldehyde/1.5–2.0 % glutaraldehyde in 0.1 M, pH 7.2 PB, is then pumped through the vasculature using the same apparatus (*see* **Notes 15** and **16**). The rinse and the first half of the fixative are run through the animal at a brisk pace to ensure that all the red blood cells are cleared. The remaining fix is then pumped through very slowly to maximize diffusion time and ensure complete tissue fixation. *N.B. Fixatives are toxic substances by definition. The perfusion should take place in a hood or on a down-draft table, and the fumes vented to the outdoors. Proper gloving, eye protection, and protective attire are a necessity.*

4. The head is removed from the body and placed in a stereotaxic head holder (Kopf). The dorsal skull is removed and the brain is blocked in the frontal plane using a micromanipulator armed with a # 22 blade attached to a pole on the micromanipulator arm. For best results when sectioning on a Vibratome, blocks should be kept at or below 1 cm in height.

5. A fiducial mark is placed in each block by driving a syringe needle through an unimportant area and leaving it there during the postfixation step [6].

6. The blocks are postfixed in the same fixative for 1–2 h at 4 °C and then stored overnight at 4 °C in 0.1 M, pH 7.2 PB.

7. Blocks are superglued to a glass plate attached to the Vibratome chuck. Then, sections are cut at 100 μm on a Vibratome (Leica VT 100S). They are collected serially and stored at 4 °C in 0.1 M, pH 7.2 PB.

3.2 Retrograde Tracing [23–26]

1. Retrograde tracer (1.0–2.0 % WGA-HRP or 1.0–2.0 % ChTB-HRP) is most conveniently injected by use of a 1.0 μl Hamilton syringe (*see* **Note 17**). For small injections, a glass micropipette can be attached to the syringe needle. Alternatively, a glass micropipette can be used to iontophoretically apply the tracer. The amount injected and the number of injections are determined by the size and shape of the target. Note that in dual tracer experiments this injection is actually made second, after the anterograde tracer injection, in a separate surgery.

2. Survival for transport of these tracers is usually 1 or 2 days, as WGA-HRP and ChTB-HRP travel by fast axoplasmic transport. The perfusion and cutting steps are described above.

3. Place sections in 0.1 M, pH 6.0 PB in a reaction tray in serial order (*see* **Note 18**). Rinse one time in 0.1 M, pH 6.0 PB is 10 min.

4. Combine molybdate-TMB reaction solutions A and B together in the dish and preincubate sections in this solution for 20 min at room temperature (~22 °C) with gentle agitation on a rotating table. Cover the reaction dish with plastic wrap to keep the alcohol from evaporating during the reaction steps.

5. Add 10 ml of 0.3 % H_2O_2 to the solution in the dish, and react for 20 min at ~22 °C with gentle agitation.

6. Repeat **step 5** two times.

7. Continue reaction overnight with gentle agitation at 4 °C in a sealed reaction dish (*see* **Note 19**).

8. Place sections in stabilizer solution, 5 % ammonium molybdate in 0.1 M, pH 6.0 PB, for 15 min with gentle agitation at ~22 °C.

9. Rinse twice in 0.1 M, pH 7.2 PB with gentle agitation at ~22 °C for 5 min (*see* **Note 20**).

10. Place in DAB protection solution for preincubation with gentle agitation at ~22 °C for 10 min. [*DAB is a carcinogen. **Steps 10–12** should be carried out in a hood.*]

11. React tissue by adding 6 ml of 0.3 % H_2O_2 to the DAB protection solution. Reaction is continued for 10–20 min with gentle agitation at ~22 °C (*see* **Note 21**). DAB *protection solution is carcinogenic and it should be properly disposed of* (*see* **Note 22**).

12. Rinse in 0.1 M, pH 7.2 PB, four times for 5 min each, with gentle agitation at ~22 °C. EM samples can be taken at this point, or the tissue can then be reacted to visualize biotinylated tracers (see below). Once EM samples have been taken, the tissue is mounted, counterstained, dehydrated, cleared, and coverslipped as a record of the areas sampled.

3.3 Anterograde Tracing [27–29]

1. Anterograde tracer (10.0 % BDA) is most conveniently injected by use of a 1.0 µl Hamilton syringe (*see* **Note 17**). For smaller injections, a glass micropipette can be attached to the syringe needle. Alternatively, a glass micropipette may be used to iontophoretically apply the tracer. The amount injected and the number of injections are determined by the size and shape of the target. Note that in dual tracer experiments this injection is made first, and the retrograde tracer injection is made in a separate surgery 1–2 days before sacrifice.

2. Survival for transport of BDA requires 7–21 days, depending on the length of the pathway being examined, as this tracer uses slow axoplasmic transport. The perfusion and cutting steps are described above.

3. Fill 12-well depression plates with 0.05 % Triton X-100 solution (500 µl per well) (*see* **Note 23**). For dual tracer studies,

transfer sections from the reaction tray into the wells. Gently agitate for 10 min at ~22 °C. Repeat two more times for a total of 30-min incubation.

4. Replace well solution with avidin-HRP solution. Gently agitate for up to 1 h at ~22 °C.

5. Continue gently agitation at 4 °C overnight in a sealed enclosure to avoid evaporation (*see* **Note 24**).

6. Transfer sections back into reaction tray and rinse in 0.1 M, pH 7.2 PB, three times for 10 min each, with gentle agitation at ~22 °C.

7. Preincubate in nickel-cobalt DAB solution for 10 min with gentle agitation at ~22 °C. *DAB is a carcinogen. Steps 7–9 should be carried out in a hood with gloves on. This solution is carcinogenic and it should be properly disposed of* (*see* **Note 22**).

8. React sections by adding 10 ml of 0.3 % H_2O_2. Reaction proceeds at ~22 °C with gentle agitation for at least 10 min, and possibly up to 30 min. The presence of cobalt chloride can produce strong background labeling (*see* **Note 25**). Some background staining is to be expected, but if the sections turn more than a light blue-gray tone, stop the reaction by proceeding to the rinse steps.

9. Rinse three times in 0.1 M, pH 7.2 PB. EM samples can be taken at this point. Once EM samples have been taken, the tissue is mounted, counterstained, dehydrated, cleared, and coverslipped as a record of the areas sampled.

3.4 Sampling, Processing, and EM Ultrathin Cutting

Tissue used for electron microscopy needs to be cut into rather small pieces to allow osmium to penetrate it. For example, we normally cut a macaque ciliary ganglion into 2–3 pieces. Cutting the brain into 100 μm sections accomplishes the same purpose. In addition, only a small sample can be cut with a diamond knife, so the area of interest is cut out with a razor blade or # 11 scalpel blade. The size of the sample is approximately 1–2 mm^2. *Remember to only work with osmium tetroxide vials in a hood. Use proper eye and hand protection. Store any solution containing any osmium tetroxide in a sealed glass container for proper disposal.*

1. Samples are collected into small glass vials filled with 0.1 M PB (pH 7.0). The following steps [2–8] include slow agitation on a rotator (EM Sciences, 65100-10).

2. Treat for 2 h with 1.0 % osmium tetroxide (OsO_4) in 0.1 M PB (pH 7.0) at ~22 °C.

3. Wash three times in dH_2O at ~22 °C.

4. Dehydrate with increasing concentrations of acetone for 10 min each (70, 90, 95 %) at ~22 °C.

5. Complete dehydration by rinsing three times in 100 % acetone at ~22 °C for 15 min each.

In the next steps the samples are embedded in epoxy resin (*see* **Note 26**).

6. Place tissue in a 1:3 mixture of Durcupan and acetone overnight at ~22 °C.

7. Place in 3:1 mixture of Durcupan and acetone for 3 h at ~22 °C.

8. Place in 100 % Durcupan for 1 h at ~22 °C.

9. Embed in fresh Durcupan (100 %) and polymerize at 60 °C overnight.

    For the actual embedding, we use either (a) conventional plastic flat-embedding molds (EM Sciences) for small pieces of tissue like ganglia or (b) plugs of resin polymerized in Beem capsules for samples taken from 100 μm thick sections. *Any leftover resin should be polymerized before disposal.*

    (a) Fill molds with fresh Durcupan until they are about half full. Place samples at the front edge of the mold and complete filling the molds with Durcupan. Try to avoid or remove bubbles. Polymerize at 60 °C overnight.

    (b) Cut lids and bottom off the Beem capsules (retain lid). Place the remaining plastic tube without bottom onto the lid and fill with 100 % Durcupan. Polymerize at 60 °C overnight. Remove and retain lid. Remove the plug from the capsule. Make sure to create your plug at least one day before embedding.

    Place a small drop of fresh, fluid 100 % Durcupan onto the cutoff lid of a Beem capsule. Position sample in that drop. Press plug into the lid, sandwiching the sample between the lid and plug. Do not squeeze the sample, as this may cause mechanical damage to the tissue. Polymerize at 60 °C. Remove and discard lid.

10. Trim blocks using a single-edged razor blade.

11. For orientation purposes, take 1.0 μm thick semithin sections of already trimmed blocks on a microtome with a glass knife. Mount them on glass slides, dry on a heating plate, and stain them for 20 s with toluidine blue. Examine sections under a light microscope to determine the area of interest. Retrim blocks to include only the area of interest.

12. Cut ultrathin silver/gold sections (~90 nm thick) with a diamond knife on an ultramicrotome (Reichert Ultracut E) and mount them onto Formvar-coated nickel slot grids (EM Sciences, G2010-Ni).

3.5 Postembedding Immunohisto-chemistry

All immunohistochemical procedures are performed on Parafilm to create a clean bench space. Attach a piece of film onto the lab bench by running the dull end of a forceps over the edges of the Parafilm. Drops (~40–50 µl) of the respective staining or washing solution for the steps described below are placed onto the Parafilm with a Pasteur pipette (see **Note 27**). Float the grids on the drops with the sample side down, contacting the fluid. We recommend that one not try to process more than four grids at the same time. All steps are done at room temperature (~22 °C).

1. Rinse grids briefly with dH_2O.

2. Etch with 3.0 % H_2O_2 in dH_2O for 1 min.

3. Wash grids with dH_2O three times for 3 min each.

4. Treat the sections with 0.1 M, pH 7.4 TBS two times for 5 and 10 min each.

5. Pretreat in TBS + 1.0 % BSA for 30 min.

6. Incubate in a solution of the primary antibody (e.g. rabbit anti-GABA IgG) in TBS + BSA + 0.05 % Tween 20 overnight. Put a glass cuvette over your floating sections and seal with Parafilm, so that the drops do not evaporate overnight.

7. Rinse sections the next day with TBS + BSA + Tween 20 three times at 10 min each.

8. Incubate in secondary antibody (goat-anti-rabbit IgG conjugated to gold particles) in TBS + BSA + Tween 20 for 2 h (see **Note 28**).

9. Rinse grids again with TBS + BSA + Tween 20 three times for 10 min each.

10. Rinse with dH_2O three times for 3 min each.

11. Fix using a solution of 2.0 % glutaraldehyde in 0.1 M, pH 7.0 EM PB for 3 min.

12. Rinse tissue with dH_2O three times for 3 min each.

3.6 Heavy Metal Staining [30]

Prepare small, clean glass vials filled with ultraclean dH_2O before starting with these steps. For four grids, you will need six vials for each washing step between and after staining with uranyl acetate and lead citrate solutions. Wash by using a forceps to move each grid gently up and down within the solution contained in a vial at least ten times per step, before advancing the grid to the next vial (see **Note 29**). The two staining steps are again done on drops of the respective solution placed on Parafilm. All steps are done at room temperature (~22 °C).

1. Stain by floating grids, section down, on a drop of 2.0 % uranyl acetate for 3 min.

2. Wash in three vials and dry as described above.

3. Stain with calcinated lead citrate solution for 30 s. Put drops of the solution next to some sodium hydroxide pellets on the Parafilm to create a dry environment. Float grids on the solution and cover with a Petri dish. Do not breathe on the sections during the staining process. The exhaled carbon dioxide will result in lead precipitation.

4. Wash in vials and dry as described above.

5. Store in grid boxes.

3.7 Analysis

The gold particles attached to the secondary antibody appear as tiny, spherical dots or freckles over the labeled elements (Fig. 1, small arrows). The uniform size and appearance of the gold particles allow them to be distinguished from ribosomes and glycogen granules. Of course, not all GABA and glycine present in cells are necessarily involved in inhibitory synaptic transmission, as these molecules have other biochemical uses. In addition, as with all immunohistochemical processes, the antibodies can attach nonspecifically. Thus, it is necessary to determine which labeling actually corresponds to inhibitory elements. One particularly effective way of confirming that a profile is positively labeled is to find the same

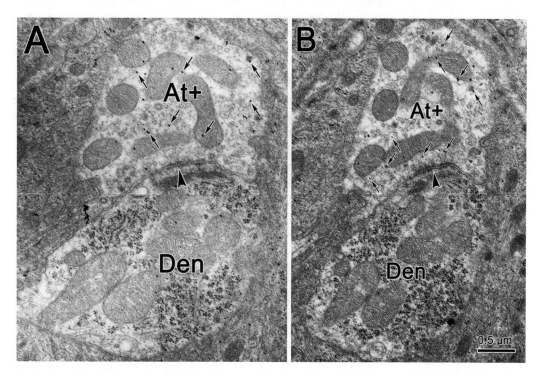

Fig. 1 Semiserial sections through a synaptic terminal in a macaque ciliary ganglion that have been labeled using postembedding immunohistochemistry for GABA. In both plates (**a**) and (**b**), numerous gold particles (*arrows*) are apparent over the terminal, indicating that it is GABA positive (At+). Synaptic density indicated by *arrowhead*. Den = dendrite. Scale in (**b**) = (**a**)

profile in semiserial sections. Background labeling should be fairly random, but if there are numerous gold particles lying over the profile in both sections, there is a high likelihood that it is labeled. Figure 1 shows such a pair, with numerous gold particles present over a GABA-positive terminal (At+) in the ciliary ganglion, but very few over other elements, such as the dendrite. Alternatively, one can quantitatively determine the background level of labeling and then establish a threshold for immunopositivity. For example, in recent studies [19, 21], we defined terminals exhibiting labeling at or below background as GABA negative and those with three times this threshold as GABA positive. Terminals with particle numbers lying between these values were placed in a non-characterized category. In this same tissue, we found that labeling in somata and dendrites was distinctly less prominent, and so we set the threshold at two times background. There are of course highly sophisticated image analysis systems for measuring particle density. However, we have found that a manual method is quite easy to use. We cut a square out of a piece of cardboard that is about the size of a typical axon terminal in our area of interest. We then slide it over the image making counts over areas that are clearly not heavily labeled. Counts from ten such areas are then averaged to define background level. Background is independently determined in this way for all the images from a single grid (see **Note 30**). This same square can then be placed over terminals, dendrites, or somata to allow them to be classified as positive or negative relative to this background level.

It is often useful to employ the postembedding immunohisto-chemical method to analyze the terminal populations contacting an identified output population. An example of this is shown in Fig. 2. The retrogradely transported tracer appears as electron dense, floc-culent, crystals within the cytoplasm of the labeled cells, and den-drites. In some cases, it will nearly fill the cytoplasm of the labeled cell, as demonstrated in the labeled dendrite (Den*) in Fig. 2a and the somata (Soma*) in Fig. 2d. In other cases, there are just small patches of electron-dense material, as shown in the somata of Fig. 2b and the dendrite of Fig. 2e. The latter is preferable when identifying synaptic contacts (arrowhead) (see **Note 31**). In Fig. 2 the GABA-positive (At+) and GABA-negative (At) terminals are easily recognized along the membranes of somatic (Fig. 2b, d) and dendritic (Fig. 2a, e) profiles. Note the association of the gold particles with mitochondria. This association is often found in GABA-stained material. Even over non-labeled profiles, gold par-ticles are generally more common over mitochondria.

Postembedding immunohistochemistry is also useful for iden-tifying whether terminals from a defined source are inhibitory. Figure 3 shows examples of terminals labeled with biotinylated dextran amine (BDA). The BDA-labeled terminals (At*) contain granular   electron-dense   reaction   product   and   are   easily

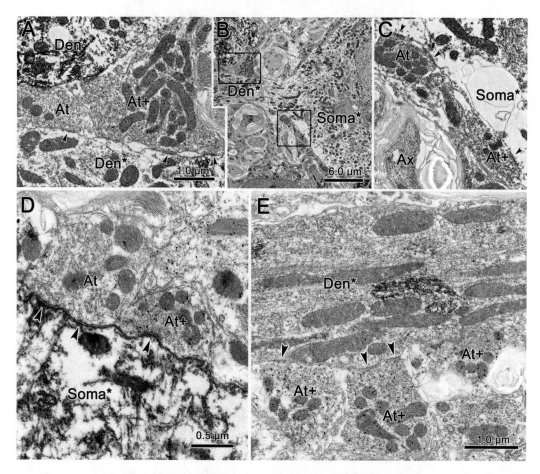

Fig. 2 Examples of combining retrograde labeling with postembedding immunohistochemistry. GABA-positive synaptic profiles (At+) are seen contacting (*arrowhead*) retrogradely labeled (ChTB-HRP) somata (Soma*) (**c, d**) and dendrites (Den*) (**a, e**). Low-magnification view in (**b**) shows sample areas for (**a**) and (**c**). Note electron-dense reaction crystals in postsynaptic elements and gold particles lying over GABA-positive terminals. At= unlabeled axon terminal. Ax=axon. Scale in (**a**)=(**c**)

discriminated from unlabeled terminals (At). In densely labeled examples (Fig. 3a, bottom and c) this product can fill the cytoplasm between the vesicles. In more lightly labeled examples (Fig. 3a, top and b) it surrounds the vesicles with an electron-dense fuzz. Those BDA-labeled terminals that are GABA positive have numerous gold particles over them (At*+) while those that are GABA negative (At*) do not (*see* **Note 32**).

Combining the anterograde and retrograde tracer techniques allow the investigator to make very forceful arguments about the input/output relationships within a nucleus. As shown in the lower magnification (a) and higher magnification (b) views in Fig. 4, both types of reaction product are present. Flocculent electron-dense crystals (arrows) are observed within this soma (Soma*) (Fig. 4a). The BDA-labeled terminal shown (At*) is completely filled with

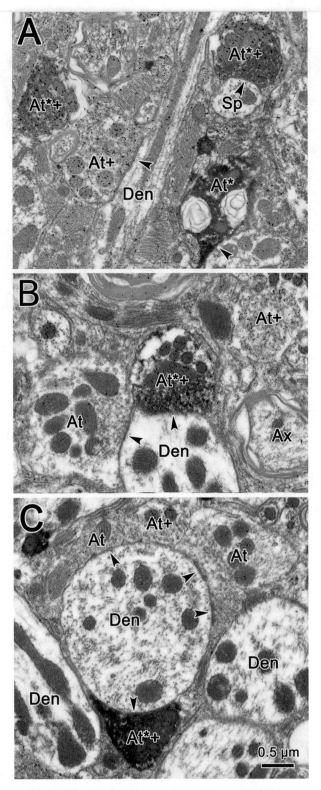

Fig. 3 Examples (**a–c**) of combining anterograde labeling with postembedding immunohistochemistry. Some GABA-positive profiles also are labeled with the anterograde tracer BDA (At*+), while others are not (At+). In addition, some BDA-labeled profiles are not GABA positive (At*). In each case, these terminals synapse (*arrowhead*) on unlabeled dendrites (Den). At=unlabeled axon terminal. Ax=axon. Scale in (**c**)=(**a**) and (**b**)

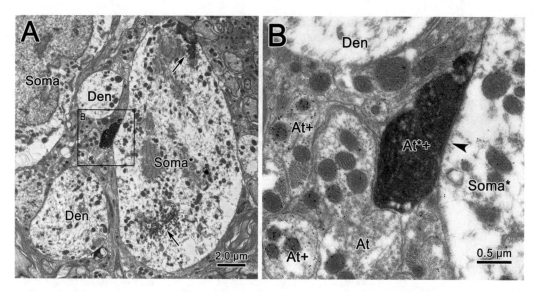

Fig. 4 Dual-tracer labeling in conjunction with postembedding immunohistochemistry. Low (**a**) and high (**b**) magnification views of a labeled synaptic contact (Box in **a**=**b**). A BDA-labeled synaptic terminal is seen contacting (*arrowhead*) a retrogradely labeled (ChTB-HRP) soma (Soma*). The densely labeled terminal also has gold particles over it, indicating that it is also GABA positive (At*+). At=unlabeled axon terminal, At+= GABA-positive axon terminal, Den=dendrite

reaction product. Nevertheless, it is still possible to visualize the gold particles indicating that this is a GABA-positive terminal (At*+), although identifying GABA-positive terminals that are not BDA labeled (At+) is considerably easier.

4   Notes

1. The solution amounts described below are for techniques developed for large feline and monkey brains. They can often be adapted for rodent work by moving the decimal points one place to the left for all of the volumes and weights.

2. Higher levels of glutaraldehyde will facilitate capture of the GABA or glycine molecules by the fix, but decrease the activity of horseradish peroxidase (HRP). Optimum fixation for HRP is with 1.25 % glutaraldehyde. So, lower levels, in the 1.5–2.0 % range, are used when doing retrograde tracing in conjunction with the postembedding immunohistochemistry. We have commonly used 1.5 % glutaraldehyde for anti-GABA experiments that are combined with retrograde tracers that employ horseradish peroxidase (HRP). We use 3.0 l of fixative for a macaque monkey perfusion and 2.0 l for a cat perfusion.

3. Do not allow the paraformaldehyde solution to heat above 60 °C. To do this, it is best to turn off the heater element

when the temperature reaches the vicinity of 50 °C, as the solution will continue absorbing heat from the hot plate. We use an Urbanti spiral ribbed funnel (Bell Art, F14648-000) and coarse #4 filter paper to speed the filtration process.

4. Since the glutaraldehyde is viscous, it tends to stick to the cylinder. This approach washes it into the fix. It is best to order the granular form of paraformaldehyde as the powder tends to become airborne and is breathed in.

5. We aliquot tracers into 0.5 ml microcentrifuge tubes (Fisher, 02-681-333) when we receive them and store them at 0 °C to maximize storage life. Once in solution, they are still fairly stable for up to a month, if refrigerated. We have described here the tracers we use most commonly, WGA-HRP, ChTB-HRP, and BDA. However, others also can be employed. Phaseolus vulgaris leucoagglutinin has also been successfully used as an anterograde tracer [21]. For short pathways, biocytin can be used instead of BDA [28, 29]. However, it should only be used with 24-h survival periods and should be injected in the same surgery as the retrograde tracer as it is quickly metabolized.

6. Of the various TMB protocols we reviewed when developing the approach, this one uses a pH closest to neutral, and it consequently is less damaging to the ultrastructure of the tissue. It also produces very little background staining and stunning signal.

7. Store TMB powder at 0 °C. Do not store solution B for more than overnight.

8. Store DAB powder at 0 °C. DAB dissolves much easier in water, than buffer, so the DAB solution is not buffered until after the DAB dissolves.

9. Triton X-100 takes a while to mix into solution even with rapid stirring. It can then be stored at 4 °C for long periods of time.

10. When received, avidin D-HRP should be stored at 4 °C.

11. The effective concentrations vary by lot. 1:250 is a good starting place, but it is likely that you will need to play with the concentration to maximize the signal-to-noise ratio for any individual antibody lot and for your tissue.

12. We have only limited experience with anti-glycine. This antibody provided us with usable data, but only at high concentrations. In the future, we plan to search for new sources and experiment with using higher glutaraldehyde concentrations in the fix.

13. Again, this is a suggested starting point for dilution. The investigator may need to play with this concentration to maximize the signal-to-noise ratio. We have also tried using 20 nm

gold particles. This makes finding the labeled terminals easier, but seems to lessen the number of gold particles present.

14. Make sure to use Millipore-filtered dH_2O for all solutions used for electron microscopy.

15. The tubing running through the pump is connected to two separate reservoirs (one for rinse and one for fix) through a Y-coupling. The tubes leading from the Y to the two reservoirs are each fitted with a quick-release hose clamp. This allows quick changeover from the rinse to the fix. The longer hose descending from the Y-connector is fitted with a stainless steel Luer-Lock hub at the open end. This is held in place with a screw-type hose clamp available from a local hardware store. One should carefully remove bubbles from the line before starting the perfusion. *Murphy's First Law of Perfusions* states: "Bubbles will inevitably lodge in the vessels leading to the area of interest, denying it fixative." The fixative is caught in a plexiglass tray. The animal lies on a perforated plexiglass platform that fits into this tray.

16. Different labs use different fixative temperatures. We are not convinced that this is a critical variable. We use room temperature wash followed by a cool fix, but this is due to the fact that the fix will cross-react if stored at room temperature. Similarly, there are many variations on the approach to the perfusion. We have come to believe that perfusions, because they have variable outcomes, engender superstitious behavior. The important variables are the animal, the skill of the experimenter, and the time before the fix hits the tissue. The latter two are obviously linked. Thus, we have directed our approach to simplifying the procedure, eliminating injections of cardiovascular agents, etc., and developing a clocklike precision to the process. We set out the instruments in order of use so that they are easy to pick up. The surgical procedure proceeds as follows:

(a) The abdomen is opened with a scalpel.

(b) The skin over the sternum is incised to meet the first incision.

(c) The sternum is split using heavy scissors.

(d) Then the diaphragm is cut with these heavy scissors.

(e) Rib spreaders are used to stabilize the ribcage.

(f) Delicate scissors are used to open the pericardium so that the heart can be slipped out.

(g) The same scissors are used to stab a small hole in the wall of the left ventricle near the apex through which the needle of the perfusion tube is then inserted.

Even with the best technique, *Murphy's Second Law of Perfusions* will come into play. It states: "The quality of the perfusion is inversely proportional to the accuracy of the tracer injection."

17. We use a blunt-tipped Hamilton syringe and sharpen the needle tip with a sharpening stone, just enough that it can penetrate the brain, but with the ejection hole near the needle tip to maintain stereotaxic accuracy. For central injections of WGA-HRP or ChTB-HRP, we use volumes on the order of 0.01–0.03 μl for central injections. Ten times this amount (0.1–0.3 μl) is used for BDA injections. ChTB-HRP is reputed to give better dendritic filling, but we do not always see this.

18. We use a plastic tackle box with a hole drilled in the base of each section and nylon mesh screening (Small Parts, CMN-2000) adhered to the bottom with fiber glass resin (Bondo) for the reaction tray. The solutions are placed in a Pyrex reaction dish that holds the tackle box. Between steps, the tackle box is blotted on squares of bench coat paper. However, for smaller sections many individuals use a piece of plexiglass that has wells drilled in it as a reaction tray and place the solutions into a staining dish.

19. Sections will be yellow at the end of the reaction step, but this does not indicate a background problem.

20. For light microscopy, the sections can be mounted out of this buffer and counterstained with cresyl violet. The reaction product is a bright blue color that can be further enhanced by using crossed polarizers.

21. Reaction product can be seen to turn from blue to brown during this period. Once this transformation has occurred the sections can be moved into the rinse.

22. Soak reaction tray and glassware in 70 % ethanol with a little Chlorox to clean reaction product from them. Rinse thoroughly with tap water followed by dH_2O.

23. Probably most labs will use plastic multiwell culture plates for this step.

24. Depression plates are placed on a wet piece of bench coat in a closed plastic box to limit evaporation. If using multiwell culture plates, employ the cover.

25. The background staining is correlated with fixation quality: the poorer the fix, the worse the background. We divide sections into three series, so if the first series stains too darkly and the background is too high, we can lower the cobalt chloride levels in a subsequent series. We usually divide sections into three series, and reserve one series for light microscopic analysis.

26. For postembedding immunohistochemical staining, the tissue needs to be embedded in a resin that allows the antibody to penetrate the tissue in order to bind to the antigen one wants to detect, but at the same time results in good ultrastructural preservation.

27. Pick up and move the grids with a nonmagnetic Dumont forceps by grabbing them along the edge. It should be noted that other labs have successfully omitted some of the steps described below. Some labs omit the etching step when they can process the tissue immediately after sectioning. The glutaraldehyde fixation step seems to help increase signal in our hands, but other labs successfully omit this step.

28. If the grid sinks into the drop, it is no longer useable since immunogold has a high affinity for the Formvar film and so will produce a high level of background staining on the back of the film, obscuring positive label on the section.

29. To make sure that the last washing step of a grid is always performed in clean, not contaminated water, start with the first grid in vial 1, proceed to vial 2, and then vial 3. Dry grid and take off the remaining fluid by touching the edge of a piece of filter paper to it. Do not touch Formvar and sections. Next rinse the second grid as previously done with grid 1, but start in vial 2, and proceed to vial 4. The third grid will begin in vial 3 and proceed to vial 5, etc. Make sure that the grid remains vertical while moving through the rinse, so that the Formvar film is not ruptured. Do not break the surface of the water except to enter and exit the vial; the surface tension of the water adds stress to the rinse steps and can rupture the Formvar film.

30. We do not make background counts over the Formvar outside the specimen, because its attractiveness to the antibody and secondary is not the same as the tissue.

31. Of course, the extent of dendritic labeling will vary with the degree of retrograde labeling overall. However, it is our impression that one can often find patches of label in small, presumably distal, dendrites at the EM level, even when these distal dendrites do not appear to be labeled at the light microscopic level. We believe that small, lightly labeled dendrites do not have enough optical density to be visualized, but this is not the case at the EM level.

32. While the gold particles are relatively easy to see when working at the EM or on a monitor, they can be difficult to demonstrate in poster or publication pictures, due to the electron-dense background of the BDA. We find that using Photoshop to lighten the image and slightly decrease the contrast can improve their visibility.

References

1. Gray EG (1969) Electron microscopy of excitatory and inhibitory synapses: a brief review. In: Akert K, Wasser PG (eds) Mechanisms of synaptic transmission, vol 31, Progress in Brain Research. Elsevier, Amsterdam, pp 141–155

2. Peters A, Palay SL, Webster HF (1991) Fine structure of the nervous system, 3rd edn. Oxford University Press, Oxford

3. Klemann CJ, Roubos EW (2011) The gray area between synapse structure and function - Gray's synapse types I and II revisited. Synapse 65:1222–1230

4. Somogyi P, Hodgson AJ, Chubb IW, Penke B, Erdei A (1985) Antisera to gamma-aminobutyric acid. II. Immunocytochemical application to the central nervous system. J Histochem Cytochem 33:240–248

5. Otterson OP (1987) Postembedding light- and electron microscopic immunocytochemistry of amino acids: descriptions of a new model system allowing identical conditions for specificity testing and tissue processing. Exp Brain Res 69:167–174

6. Charara A, Smith Y, Parent A (1996) Glutamatergic inputs from the pedunculopontine nucleus to midbrain dopaminergic neurons in primates: *Phaseolus vulgaris*-leucoagglutinin anterograde labeling combined with postembedding glutamate and GABA immunohistochemistry. J Comp Neurol 364:254–266

7. Chun MH, Wassle H (1989) GABA-like immunoreactivity in the cat retina: electron microscopy. J Comp Neurol 279:55–67

8. Decavel C, Van Den Pol AN (1990) GABA: a dominant neurotransmitter in the hypothalamus. J Comp Neurol 302:1019–1037

9. Todd AJ (1991) Immunohistochemical evidence that acetylcholine and glycine exist in different populations of GABAergic neurons in lamina III of rat spinal dorsal horn. Neuroscience 44:741–746

10. Oliver DL, Winer JA, Beckius GE, Saint Marie RL (1994) Morphology of GABAergic neurons in the inferior colliculus of the cat. J Comp Neurol 340:27–42

11. Mize RR, Whitworth RH, Nunes-Cardoza B, Van Der Want J (1994) Ultrastructural organization of GABA in rabbit superior colliculus revealed by quantitative postembedding immunocytochemistry. J Comp Neurol 341:273–287

12. Kawaguchi K, Kubota Y (1997) GABAergic cell types and their synaptic connections in rat frontal cortex. Cereb Cortex 7:476–486

13. Van Horn SC, Ersir A, Sherman SM (2000) Relative distribution of synapses in the A-lamina of the lateral geniculate nucleus. J Comp Neurol 416:509–520

14. Knott GW, Quairiaux C, Genoud C, Welker E (2002) Formation of dendritic spines with GABAergic synapses induced by whisker stimulation in adult mice. Neuron 34:265–273

15. Thind KK, Yamawaki R, Phanwar I, Zhang G, Wen X, Buckmaster PS (2010) Initial loss but later excess of GABAergic synapses with dentate granule cells in a rat model of temporal lobe epilepsy. J Comp Neurol 518:647–667

16. Smith Y, Bolam JP (1990) The output neurones and the dopaminergic neurones of the substantia nigra receive a GABA-containing input from the globus pallidus in the rat. J Comp Neurol 296:47–64

17. de la Roza C, Reinoso-Suarez F (2006) GABAergic structures in the ventral part of the oral pontine reticular nucleus: an ultrastructural study. Neuroscience 142:1183–119318

18. Omelchemko N, Bell R, Sesack SR (2009) Lateral habenula projections to dopamine and GABA neurons in the rat ventral tegmental area. Eur J Neurosci 30:1239–1250

19. Wang N, Warren S, May PJ (2010) The macaque midbrain reticular formation sends side-specific feedback to the superior colliculus. Exp Brain Res 201:701–717

20. Day-Brown JD, Wei H, Chomsung RD, Petry HM, Bickford ME (2010) Pulvinar projections to the striatum and amygdala in the tree shrew. Front Neuroanat 4:143. doi:10.3389/fnana.2010.0014

21. Wang N, Perkins E, Zhou L, Warren S, May PJ (2013) Anatomical evidence that the superior colliculus controls saccades through central mesencephalic reticular formation gating of omnipause neuron activity. J Neurosci 33:16285–16296

22. Nakamota KT, Mellott JG, Killius J, Storey-Workley ME, Sowick CS, Schofield BR (2013) Ultrastructural examination of the corticocollicular pathway in the guinea pig: a study using electron microscopy, neural tracers, and GABA immunocytochemistry. Front Neuroanat 7:13. doi:10.3389/fnana.2013.00013

23. Olucha F, Martínez-García F, López-García C (1985) A new stabilizing agent for the tetramethyl benzidine (TMB) reaction product in the histochemical detection of horseradish peroxidase (HRP). J Neurosci Methods 13:131–138

24. Chen B, May PJ (2002) Premotor control of eyelid movements in conjunction with vertical saccades in the cat: the rostral interstitial nucleus of the medial longitudinal fasciculus. J Comp Neurol 450:183–202

25. Perkins E, Warren S, Lin RC-S, May PJ (2006) Projections of somatosensory cortex and frontal eye fields upon incertotectal neurons in the cat. Anat Rec 288A:1310–1329

26. Adams JC (1977) Technical considerations on the use of horseradish peroxidase as a neuronal marker. Neuroscience 2:141–145

27. Somgyi J (2002) Differences in ratios of GABA, glycine and glutamate immunoreactivities in nerve terminals on rat hindlimb motoneurons: a possible source of post-synaptic variability. Brain Res Bull 59:151

28. Sun W, May PJ (2014) Central pupillary light reflex circuits in the cat: the olivary pretectal nucleus. J Comp Neurol 522:3960–3977

29. Sun W, May PJ (2014) Central pupillary light reflex circuits in the cat: morphology, ultrastructure and inputs of preganglionic motoneurons. J Comp Neurol 522:3978–4002

30. Hanaichi T, Sat T, Iwamoto T, Malavasi-Yamashiro J, Hoshino M, Mizuno M (1986) A stable lead by modification of Sato's method. J Electron Microsc (Tokyo) 35:304–306

Neuromethods (2016) 115: 205–216
DOI 10.1007/7657_2015_74
© Springer Science+Business Media New York 2015
Published online: 18 July 2015

Molecular Electron Microscopy in Neuroscience: An Approach to Study Macromolecular Assemblies

Bjoern Sander and Monika M. Golas

Abstract

Macromolecular assemblies play a critical role in the function of the nervous system. They contribute to a multitude of cellular processes such as cell type-specific gene expression, intracellular trafficking, and synaptic transmission. Macromolecular assemblies are composed of a number of proteins and can likewise comprise additional molecules such as DNA or RNA. However, these complexes are typically dynamic entities that can change their conformation and composition dependent on their functional state. The visualization of these macromolecular complexes and assembly intermediates by transmission electron microscopy (EM) techniques can provide insights into their function in the cell as well as uncover their molecular mechanisms and life cycle. In this chapter, we describe how to prepare and analyze macromolecular assemblies by conventional negative staining EM, negative staining cryo-EM, and unstained cryo-EM. In particular, we focus on methods that allow imaging macromolecular assemblies that are near the size or concentration limits of the method or that are only transiently formed. Such challenging macromolecular assemblies can be subjected to the GraFix approach that cross-links the complexes under ultracentrifugation. Moreover, we also discuss EM imaging techniques and image processing approaches suited to compare macromolecular complexes captured at different conformational or compositional states by using difference mapping. Together, this set of single-particle EM methods provides an outline of how macromolecular complexes of the nervous system can be studied by EM techniques.

Keywords Electron cryomicroscopy (cryo-EM), Macromolecular assembly, GraFix, Negative staining, Neuroscience, Single-particle electron microscopy, Image processing, Difference mapping, 3D structure

1 Introduction

Cellular processes essential for the function of the nervous system are carried out by a multitude of macromolecular assemblies that amongst others contribute to cell type-specific gene expression, intracellular trafficking, and synaptic transmission [1–4]. These macromolecular assemblies are composed of multiple factors such as proteins and nucleic acids. Typically, the complexes undergo conformational or compositional changes essential for the functional activity of the assemblies [5]. In particular, factors can be recruited or destabilized, and the complexes can adopt different conformations.

Transmission electron microscopy (EM) is the method of choice to visualize complex assemblies [6, 7]. EM methods can be subdivided into cellular and molecular EM approaches. Cellular EM aims at describing the ultrastructure of tissues and cultured cells using ultrathin sections of the specimen [8]. Sections can be immuno-gold labeled or labeled with streptavidin-gold or similar systems in order to localize a factor of interest [9]. In neuroscience, cellular EM has amongst others contributed to the description of synapses [9, 10]. Molecular EM comes in three different flavors [6]: (1) electron tomography, (2) electron crystallography, and (3) single-particle EM. All of these methods aim at describing the structural organization of macromolecules. Of note, molecular EM can not only describe the overall structural organization of the assemblies, but can also provide insights into their dynamical changes when complexes captured at different functional states are compared [11].

Electron tomography is used to study isolated macromolecules, organelles, cells, or sections thereof, thus being at the interface between cellular and molecular EM. Objects are imaged under different tilt angles, and a three-dimensional reconstruction is determined from the tilt series. In contrast to the two other types of molecular EM methods, electron tomography can also be performed on objects with a unique structure, and it has, e.g., been used to study synaptic vesicles [12] as well as the rod sensory cilium [13]. In case a macromolecule can be crystallized as two-dimensional (2D) sheets or helical tubes, electron crystallography can be used to study the structure of the macromolecule. Amongst others, electron crystallography has been used to determine the structure of the acetylcholine receptor [14]. Finally, in single-particle EM, purified macromolecules are imaged under different projection directions without the need to crystallize the particles. This approach has been used to study, e.g., the gamma-secretase [15]. The projection images taken in the electron microscope are used to determine the 2D and three-dimensional (3D) structure of a given macromolecule. In its original design, the approach requires that the images represent particles with the same conformation and composition, although it more and more becomes possible to deal with increasing levels of computational complexity due to heterogeneity [16, 17].

Here, we describe approaches used to visualize and analyze macromolecular assemblies by single-particle EM. In particular, we describe the GraFix protocol [18] which offers advantages when dealing with particles at a low concentration (<50 μg/ml), small particles (<250 kDa), or fragile particles that tend to disintegrate during sample preparation. Moreover, we discuss the sandwich negative staining method, which can be used at room temperature and under cryogenic conditions [19] as well as unstained cryo-EM by freeze-plunging into liquid ethane [20].

Finally, we describe how difference mapping can be used to identify changes of the particle's architecture induced by, e.g., the binding or dissociation of protein factors [11, 21].

In this chapter, we focus on the survival of motor neuron (SMN) complex that represents a prototype of a dynamic macromolecular complex. Mutations of the *SMN1* gene can result in spinal muscular atrophy, a severe disorder that results in a progressive degeneration of spinal motor neurons [3, 22]. A role of SMN in regulating the localization of poly(A) mRNA in motor neurons has been described [23]. Moreover, the SMN complex assists in the assembly of spliceosomal small nuclear ribonucleoproteins (snRNPs) [24] restricting the deposition of the hetero-heptameric, ring-forming Sm core on correct target RNA only [25]. To this end, the SMN complex collaborates with pICln that acts as a molecular chaperone preventing the premature binding of RNA [11]. The SMN complex ejects pICln from the partially preassembled Sm core complex, thus enabling the binding of the correct RNA target [11].

2 Materials

2.1 GraFix Sample Preparation

1. Purified macromolecular complex in the buffer of choice (*see* **Note 1**).

2. Primary amine-free gradient buffer, e.g., 5 % gradient buffer (20 mM HEPES pH 7.5, 150 mM NaCl, 5 % (v/v) glycerol), 20 % gradient buffer (20 mM HEPES pH 7.5, 150 mM NaCl, 20 % (v/v) glycerol, 0.1 % (v/v) glutaraldehyde). Use EM-grade glutaraldehyde (25 % w/v solution) and prepare buffer freshly before use. (CAUTION: Glutaraldehyde is toxic and requires appropriate handling.) *See* **Note 2**.

3. Gradient Master gradient forming device (Biocomp, Fredericton, NB, Canada).

4. Swing-out ultracentrifugation rotor TH-660 (SORVALL®) or SW60Ti (Beckman®) or similar with suitable centrifugation tubes. *See* **Note 3**.

5. SORVALL®, Beckman®, or similar ultracentrifuge.

6. Fractionation device: Tube/needle system connected to a peristaltic pump (P1, GE Healthcare®) or tube-piercing device (Brandel Isco tube piercer, Isco, Lincoln, USA).

2.2 Sandwich Carbon Negative Staining and Cryo Negative Staining

1. 2 % (w/v) uranyl formate: Prepare freshly, protect from light, and keep at 4 °C during specimen preparation. (CAUTION: Follow strictly the safety instructions valid in your country.)

2. Copper EM grids covered with a holey carbon film, either homemade or commercial (Quantifoil®, Quantifoil Micro

Tools, Jena, Germany, or C-flat, Protochips, Raleigh, NC, USA).

3. Homemade continuous carbon film prepared by evaporation of carbon on freshly cleaved mica (indirect coating).

4. Workshop-made black Teflon® or black plastic block with holes of about 25 and 150 µl. *See* **Note 4**.

5. Cryo negative staining only: Liquid nitrogen and liquid nitrogen container for storage of EM grids. (CAUTION: Use appropriate safety precautions when handling liquid nitrogen. Liquid nitrogen can cause severe burns and asphyxia).

2.3 Unstained Cryo-EM on Carbon Film

1. For buffer exchange only: Zeba spin columns (Pierce, Rockford, IL, USA).

2. Copper EM grids covered with a holey carbon film.

3. Homemade continuous carbon film on mica.

4. Workshop-made black Teflon or black plastic block. *See* **Note 4**.

5. Liquid nitrogen, liquid nitrogen container for storage of EM grids, and ethane gas. (CAUTION: Liquid nitrogen and ethane can cause severe burns and asphyxia. Ethane is extremely flammable and forms explosive mixtures with air. Follow appropriate safety precautions.)

6. Vitrobot (FEI®, Eindhoven, The Netherlands), EM CPC (Leica®, Wetzlar, Germany), or similar freeze-plunging device.

7. Glow-discharging device (e.g., CTA 005, Bal-TEC, Liechtenstein or similar).

2.4 Imaging of the EM Specimens

1. Electron microscope (e.g. FEI® or Jeol®, Tokyo, Japan, or similar) equipped with a side-entry specimen holder (Gatan®, Pleasanton, CA, USA) or autoloader system (FEI®) and an imaging system (photographic film, CCD, direct electron detector). Cryo-shields, cryobox, or similar required for cryo-EM work.

2. For imaging on photographic film only: High-resolution slide scanner.

2.5 Difference Mapping

1. Image processing software for single-particle EM, e.g., IMAGIC (https://www.imagescience.de/imagic.html), SPIDER (http://spider.wadsworth.org/spider_doc/spider/docs/spider.html), EMAN (http://ncmi.bcm.edu/ncmi/software/software_details?selected_software=counter_222), FREALIGN (http://grigoriefflab.janelia.org/frealign), or XMIPP (http://xmipp.cnb.csic.es/twiki/bin/view/Xmipp/WebHome).

2. R Statistical Computing package (http://www.r-project.org/).

3 Methods

3.1 GraFix Sample Preparation

1. Purify the macromolecular assembly of interest according to your established protocol and check by appropriate biochemical methods such as SDS-PAGE, Western blotting, mass spectrometry, and functional assays (see, e.g., [11] for the purification and characterization of the SMN-related complexes).

2. Prepare the gradient in the centrifugation tubes (*see* Fig. 1a, b): Mark the required filling height for the high-percentage gradient buffer using the tube height labeling tool of the Gradient Master. Fill the centrifugation tube with precooled low-percentage gradient buffer (i.e., the 5 % gradient buffer for the SMN subcomplexes) to a level slightly above the mark. Thereafter, fill the high-percentage gradient buffer (i.e., the 20 % gradient buffer for the SMN subcomplexes) to the bottom part of the centrifugation tube using a syringe and a long needle. Fill up to the mark. Cap and run gradient forming program. Keep gradients at 4 °C for about 1 h before loading the sample. *See* **Note 5**.

3. Load the sample either directly on the gradient (if no primary amines are present in the buffer) or pipet a cushion of a primary amine-free buffer (200 µl in case of a TH-660 gradient) on top of the gradient followed by the sample (see Fig. 1c). Be careful not to disturb the gradient during loading (*see* **Note 5**). Load at least 10–20 µg of sample (preferably: 50–500 µg; *see* **Note 6**).

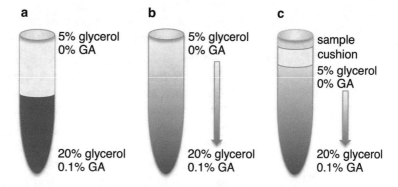

Fig. 1 Preparation of the GraFix gradient. (**a**) The lower percentage gradient solution is filled into the centrifuge tube, and thereafter, the high-percentage gradient solution is filled using a syringe/needle. *GA* glutaraldehyde. (**b**) After tilting and rotating, a gradient is formed which can directly be used to load the sample in case no primary amines are used in the sample buffer. (**c**) In case of primary amines in the sample buffer (e.g., elution peptides), a cushion comprising a primary amine-free buffer can be pipetted on the gradient before sample loading

4. Run the gradient in the ultracentrifuge at 4 °C. For the SMN subcomplexes, we recommend the following conditions: TH-660, 5–20 % glycerol gradient, 35,000–40,000 rpm at 18 h and 4 °C (*see* **Note 7**).

5. Fractionate the gradient from the bottom using a tube/needle system connected to peristaltic pump or a tube-piercing system (*see* **Note 8**).

6. Identify peak fractions by measuring the absorbance at 280 nm (*see* **Note 9**).

3.2 Sandwich Carbon Negative Staining and Cryo Negative Staining

1. Fill the sample (GraFix fraction or eluate) into the small sample wells (ca. 25 μl) as well as the staining solution into the larger wells (ca. 150 μl) of the sample block. *See* **Note 10**. Optionally, a washing buffer can also be filled into one of the large wells. Cool the preparation block with ice.

2. Cut a ca. 3 × 3 mm piece of continuous carbon film on mica and place it carefully into the sample solution (*see* **Note 11**). The carbon film should float on the particle solution, but not completely detach from the mica.

3. Incubate the carbon film floating on the sample for a defined period. A wide range from 15 s to 48 h can be used dependent on the concentration and adsorption properties of the sample. For the SMN-related complexes, we typically used an incubation time of 15–90 s. Do not interrupt the cooling chain during incubation.

4. Optional washing step (*see* **Note 12**): Take out the carbon film with the mica support from the sample solution, blot carefully with filter paper, and transfer to the well filled with washing buffer so that the carbon film floats on top of the washing buffer. Incubate for 15–30 s. This step can be repeated, if needed.

5. Take out the carbon film/mica support, blot carefully with filter paper, and transfer to the staining solution. The carbon film should now completely detach from the mica support.

6. Lift the carbon film out of the staining solution by placing an EM grid on top of the carbon film. Blot carefully from the side without damaging the carbon film.

7. Float another piece of carbon film on the surface of the second well filled with staining solution. The mica should completely detach from the carbon film. Submerge the EM grid underneath the floating carbon film and form the sandwich by lifting the grid out of the staining solution. Blot carefully using filter paper.

8. For room-temperature negative staining, let the EM grid air-dry and store it at a dry, dark place at ambient temperature. For cryo negative staining, freeze grid in liquid nitrogen and store in liquid nitrogen.

3.3 Unstained Cryo-EM on Carbon Film

1. Optional step: Exchange the buffer to a cryo-EM compatible buffer using a spin column. *See* **Note 13**.

2. Fill the sample into the ca. 25 μl sample wells of the precooled, black preparation block and cool the block with ice.

3. Adsorb the particles on a small piece of mica (ca. 3 × 3 mm) by floating the carbon film on the surface of the particle solution. The mica should completely detach from the carbon film.

4. Incubate the carbon film floating on the sample for a defined period of time depending on properties of the sample. Do not interrupt the cooling chain during incubation.

5. Lift the carbon film out of the particle solution by placing an EM grid on top of the carbon film.

6. Mount quickly in the freeze-plunging device and pipet 4–5 μl of buffer on the grid (*see* **Note 14**). Do not let the EM grid dry.

7. Blot and plunge into liquid ethane.

8. Transfer into liquid nitrogen and keep in liquid nitrogen.

3.4 Imaging of the EM Specimens

1. Mount the sample in an appropriate specimen holder and transfer into the electron microscope. Keep the specimen under cryogenic conditions in case of cryo-EM specimens.

2. Image the sample in the electron microscope. For the SMN-related subcomplexes, we use a magnification of 343,000–380,000 and 4× binning in an electron microscope equipped with a charge-coupled device (CCD) camera with 15 μm pixel size [26]. On other imaging devices, magnifications of about 50,000–100,000× will lead to comparable results [27]. Example raw EM images showing the 6S pICln/SmD1/SmD2/SmE/SmF/SmG complex and the 8S pICln/SMNΔC/Gemin2/SmD1/SmD2/SmE/SmF/SmG transfer intermediate complex are shown in Fig. 2.

3.5 Difference Mapping

1. Pick individual macromolecules from the digital EM images. Use manual selection protocols or (semi-) automated particle selection software as included in all major single-particle image processing packages.

2. Correct for the contrast-transfer function (CTF) using CTF correction software (e.g., [28–32]).

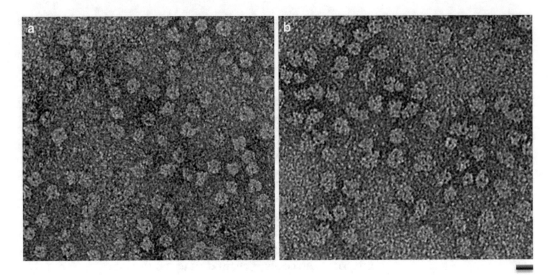

Fig. 2 Raw EM images of SMN-related complexes. Purified complexes were subjected to sandwich negative staining and imaged in an electron microscope. Scale bar, 10 nm. (**a**) Raw EM image of the 6S pICln/SmD1/ SmD2/SmE/SmF/SmG complex. (**b**) Raw EM image of the 8S pICln/SMNΔC/Gemin2/SmD1/SmD2/SmE/SmF/ SmG transfer intermediate

3. Filter particle images and align the individual particle images either against the sum of all images or some selected individual particle images using one of the abovementioned image processing tools or individual programs [29–33]. Classify using an unsupervised clustering algorithm and generate "class average" images by averaging each class of images (*see* **Note 15**).

4. Use selected class averages as references for a new round of alignment. Repeat classification and alignment until the result is stable.

5. For difference mapping, band-pass filter the class averages to the same resolution, normalize (if not included in the filtering algorithm), and align the images of one data set using the class averages of the second data set as references.

6. Identify class averages imaged under a similar projection angle and subtract class averages from each other. Visualize the difference map by color-coding according to the standard deviation. To indicate the position of the class average in the difference maps, a contour line can be manually drawn around the smaller of the two particles and transferred to the same position in the difference map using a standard vector graphics program (*see* Fig. 3).

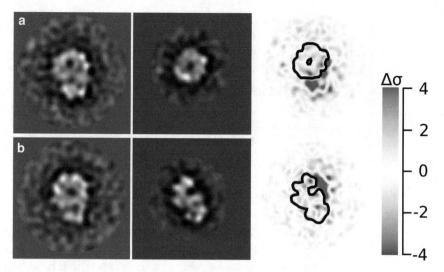

Fig. 3 Difference mapping of SMN-related complexes. Shown are the selected, filtered class averages (*left* and *middle*) as well as their difference map (*right*; see color scale ranging from -4σ to 4σ) including a contour line for the smaller of the two complexes [11]. (**a**) Comparison of the 8S plCln/SMNΔC/Gemin2/SmD1/SmD2/SmE/SmF/SmG transfer intermediate complex and the 6S plCln/SmD1/SmD2/SmE/SmF/SmG complex. (**b**) Comparison the 8S plCln/SMNΔC/Gemin2/SmD1/SmD2/SmE/SmF/SmG transfer intermediate complex and the 7S SMNΔC/Gemin2/SmD1/SmD2/SmE/SmF/SmG complex

4 Notes

1. If glycerol, sucrose, or similar carbohydrates are used during purification, the concentration of these substances must be at least 2–3 % lower (better at least 5 %) than their concentration in the low-percentage gradient buffer; for example, use max. 2–3 % glycerol in the sample in case of a 5–20 % gradient. This is essential to prevent that the sample sinks into the gradient during sample loading. If in doubt, the sinking behavior can be checked by pipetting a few ml of the lower percentage gradient buffer into a 15 ml centrifuge tube onto which a small aliquot of the sample is slowly pipetted. If the sample sinks, dilute the sample with carbohydrate-free buffer and repeat sinking test until the sample stays on top of the gradient buffer.

2. In case the sample buffer contains a primary amine (e.g., elution peptide), a primary amine-free cushion should be pipetted on the gradient before loading the sample. This cushion prevents the direct contact of low-molecular-weight substances in the sample with the cross-linker and is usually superior to, e.g., buffer exchange or dialysis procedures. Note that the latter procedures may lead to unwanted loss or degradation of fragile samples.

3. For macromolecular assemblies where subcomplexes or contaminants are co-purified with the complex of interest, the usage of a longer centrifugation tube (volume ca. 13.2 ml) is

recommended (e.g., SORVALL® TH-641 rotor or Beckman® SW41Ti rotor). For highly pure samples, we recommend a shorter centrifugation tube (volume ca. 4.2 ml; e.g., SOR-VALL® TH-660 or Beckman® SW60Ti).

4. We recommend a black block as the black color is favorable in order to see the floating carbon film. Use preferentially material for the preparation block with low affinity for proteins in order to minimize sample loss.

5. Gradients have to be prepared and transported with care to avoid unwanted mixing. Observe the phase boundary during filling of the high-percentage buffer. Do not introduce air bubbles.

6. The optimum loading amount is about 100–200 µg of sample dependent on the sample properties. Lower sample amounts can be used; however, a prolonged adsorption period will be required to increase particle numbers on the carbon film. We do not recommend to use sample amounts >500 µg as this can lead to inter-particle cross-linking. The sample volume loaded should not exceed 10 % of the gradient volume to ensure a good separation during centrifugation.

7. Centrifugation conditions should be optimized so that the macromolecular assembly of interest runs to about 50–85 % of the gradient length. This ensures that low-molecular-weight substances that can decrease the image contrast are well separated from the sample.

8. Avoid perturbing of the gradient during fractionation. Take special care when introducing the needle into the gradient, and make sure that the needle is placed at the very bottom of the gradient tube. Alternatively, fractionation devices (see above) that perforate the bottom of the tube may be used.

9. An optional parallel non-GraFix gradient can be used for bio-chemical analysis of the fractions by SDS-PAGE, Western blotting, mass spectrometry, and/or functional assays to confirm the identity of the particle. It should be noted, however, that some fragile assemblies are disassociated during ultracentrifugation in non-GraFix gradients resulting in a different migration behavior in the gradients or lower yield [19, 21]. Therefore, the gradient profiles of the GraFix and the non-GraFix fractions should be checked. Note that the constituents of assemblies treated by GraFix cannot be separated by SDS-PAGE due to the cross-linking; however, fractions can none-theless be analyzed by mass spectrometry [34].

10. We recommend freshly prepared uranyl formate solutions (2 % in water). Other solutions such as uranyl acetate, phospho-tungstic acid (PTA), or ammonium molybdate are used for staining of single particles as well. It can be advantageous to compare different stains and select the one that results in the

highest contrast and no/least artifacts (e.g., pH-induced aggregation of sample).

11. We recommend preparing continuous carbon film of different thickness and testing for each sample which carbon film works best. Carbon films should be stored at a clean, dry, and dark place.

12. The washing step is recommended in case the contrast of the sample in the electron microscope is low or in case staining artifacts such as a phenomenon commonly called "champagning" is observed.

13. Buffer exchange is required in case the sample contains a higher concentration of glycerol, sucrose, or another substance reducing the image contrast of cryo-EM images or interfering with the vitrification process. This applies, e.g., for GraFix gradient fractions.

14. In case of a low sample concentration, sample solution can be used instead of buffer.

15. We typically use class sizes of ca. 10–40 individual particle images in average per class.

Acknowledgements

This work was supported by a Lundbeck Foundation's Fellowship, the Sapere Aude Program of the Danish Council for Independent Research, the Novo Nordisk Foundation, A. P. Møller Foundation for the Advancement of Medical Sciences, and the Danish Center for Scientific Computing to MMG as well as Danish Council for Independent Research, A. P. Møller Foundation for the Advancement of Medical Sciences, and the Novo Nordisk Foundation to BS.

References

1. Haucke V et al (2011) Protein scaffolds in the coupling of synaptic exocytosis and endocytosis. Nat Rev Neurosci 12:127–138

2. Millecamps S, Julien JP (2013) Axonal transport deficits and neurodegenerative diseases. Nat Rev Neurosci 14:161–176

3. Burghes AH, Beattie CE (2009) Spinal muscular atrophy: why do low levels of survival motor neuron protein make motor neurons sick? Nat Rev Neurosci 10:597–609

4. Yoo AS, Crabtree GR (2009) ATP-dependent chromatin remodeling in neural development. Curr Opin Neurobiol 19:120–126

5. Jahn R, Fasshauer D (2012) Molecular machines governing exocytosis of synaptic vesicles. Nature 490:201–207

6. Sander B, Golas MM (2011) Visualization of bionanostructures using transmission electron microscopical techniques. Microsc Res Tech 74:642–663

7. Milne JL et al (2013) Cryo-electron microscopy – a primer for the non-microscopist. FEBS J 280:28–45

8. Banks RW (1999) Cytological staining methods. In: Windhorst U, Johansson H (eds) Modern techniques in neuroscience research. Springer, Berlin, pp 1–26

9. Harris KM, Weinberg RJ (2012) Ultrastructure of synapses in the mammalian brain. Cold Spring Harb Perspect Biol 4(5). pii: a005587

10. Torrealba F, Carrasco MA (2004) A review on electron microscopy and neurotransmitter systems. Brain Res Brain Res Rev 47:5–17

11. Chari A et al (2008) An assembly chaperone collaborates with the SMN complex to generate spliceosomal SnRNPs. Cell 135:497–509

12. Fernandez-Busnadiego R et al (2013) Cryo-electron tomography reveals a critical role of RIM1alpha in synaptic vesicle tethering. J Cell Biol 201:725–740

13. Gilliam JC et al (2012) Three-dimensional architecture of the rod sensory cilium and its disruption in retinal neurodegeneration. Cell 151:1029–1041

14. Miyazawa A et al (2003) Structure and gating mechanism of the acetylcholine receptor pore. Nature 423:949–955

15. Renzi F et al (2011) Structure of gamma-secretase and its trimeric pre-activation intermediate by single-particle electron microscopy. J Biol Chem 286:21440–21449

16. Sander B et al (2010) An approach for de novo structure determination of dynamic molecular assemblies by electron cryomicroscopy. Structure 18:667–676

17. Scheres SH et al (2009) Maximum likelihood refinement of electron microscopy data with normalization errors. J Struct Biol 166:234–240

18. Kastner B et al (2008) GraFix: sample preparation for single-particle electron cryomicroscopy. Nat Methods 5:53–55

19. Golas MM et al (2009) Snapshots of the RNA editing machine in trypanosomes captured at different assembly stages in vivo. EMBO J 28:766–778

20. Adrian M et al (1984) Cryo-electron microscopy of viruses. Nature 308:32–36

21. Golas MM et al (2010) 3D cryo-EM structure of an active step I spliceosome and localization of its catalytic core. Mol Cell 40:927–938

22. Lunn MR, Wang CH (2008) Spinal muscular atrophy. Lancet 371:2120–2133

23. Fallini C et al (2011) The survival of motor neuron (SMN) protein interacts with the mRNA-binding protein HuD and regulates localization of poly(A) mRNA in primary motor neuron axons. J Neurosci 31:3914–3925

24. Meister G et al (2001) A multiprotein complex mediates the ATP-dependent assembly of spliceosomal U snRNPs. Nat Cell Biol 3:945–949

25. Pellizzoni L et al (2002) Essential role for the SMN complex in the specificity of snRNP assembly. Science 298:1775–1779

26. Sander B et al (2005) Advantages of CCD detectors for de novo three-dimensional structure determination in single-particle electron microscopy. J Struct Biol 151:92–105

27. McMullan G et al (2009) Detective quantum efficiency of electron area detectors in electron microscopy. Ultramicroscopy 109:1126–1143

28. Sander B et al (2003) Automatic CTF correction for single particles based upon multivariate statistical analysis of individual power spectra. J Struct Biol 142:392–401

29. van Heel M et al (1996) A new generation of the IMAGIC image processing system. J Struct Biol 116:17–24

30. Shaikh TR et al (2008) SPIDER image processing for single-particle reconstruction of biological macromolecules from electron micrographs. Nat Protoc 3:1941–1974

31. Tang G et al (2007) EMAN2: an extensible image processing suite for electron microscopy. J Struct Biol 157:38–46

32. Sorzano CO et al (2004) XMIPP: a new generation of an open-source image processing package for electron microscopy. J Struct Biol 148:194–204

33. Sander B et al (2003) Corrim-based alignment for improved speed in single-particle image processing. J Struct Biol 143:219–228

34. Richter FM et al (2010) Merging molecular electron microscopy and mass spectrometry by carbon film-assisted endoproteinase digestion. Mol Cell Proteomics 9:1729–1741

Neuromethods (2016) 115: 217–243
DOI 10.1007/7657_2015_96
© Springer Science+Business Media New York 2016
Published online: 23 February 2016

Assessment of Apoptosis and Neuronal Loss in Animal Models of HIV-1-Associated Neurocognitive Disorders

Jean-Pierre Louboutin, Beverly Reyes, Lokesh Agrawal, Elisabeth Van Bockstaele, and David S. Strayer

Abstract

HIV-1-associated neurocognitive disorder (HAND) is a neurodegenerative disease resulting in various clinical manifestations, characterized by neuroinflammation, oxidative stress, and related events. Neuronal damage in HAND is felt to be mainly indirect: microglial cells infected by HIV-1 increase the production of cytokines and release HIV-1 proteins, the most likely neurotoxins, among which are the envelope proteins gp120 and gp41 and the nonstructural proteins Nef, Rev, Vpr, and Tat. We review and discuss here different methods used in the assessment of apoptosis and neuronal loss in different experimental, acute and chronic, models of HAND. We also briefly consider how these techniques help to evaluate the effects of gene delivery of antioxidant enzymes in animal models of HAND.

Keywords: Immunohistochemistry, HIV-1, Apoptosis, Neuronal loss, Caspase, Gene therapy, Oxidative stress, gp120, Tat, HIV-1-associated neurocognitive disorder, Transmission electron microscopy, Morphometry, Cell culture

1 Background and Historical Review

1.1 General Considerations

We describe here the application of a number of methods to characterize apoptosis and neuronal death in experimental models of HIV-1-associated neurocognitive disorder (HAND).

HIV-1 enters the central nervous system (CNS) soon after it enters the body. Advances in the treatment of HIV-1 have dramatically improved survival rates over the past 10 years, but HIV-associated neurocognitive disorders (HAND) remain highly prevalent and continue to represent a significant public health problem, partly because highly active antiretroviral therapeutic drugs (HAART) penetrate the CNS poorly. In the early 1990s, the neurologic complications of HIV-1 infection were classified into two levels of disturbance: (1) HIV-associated dementia (HAD) and (2) minor cognitive motor disorder (MCMD) [1, 2]. HAD was considered as the most common cause of dementia in adults under 40 (associating subcortical dementia, with basal ganglia involvement, manifesting as psychomotor slowing, parkinsonism, behavioral abnormalities, and cognitive difficulties) [1–3], but HAD has become less common since HAART was introduced [2]. This reduction probably reflects better control of HIV in the periphery, since antiretroviral drugs penetrate the CNS poorly. MCMD described a less severe presentation of HIV-associated neurocognitive impairment that did not meet criteria for HAD.

More recently, in light of the changing epidemiology of HIV infection, particularly the increase in risk factors and comorbid associations not previously described, including substance abuse disorders (e.g., methamphetamine dependence), medical conditions associated with HAART treatment (e.g., hyperlipidemia), and comorbid infectious diseases (e.g., hepatitis C virus), the diagnostic criteria for HAND have been updated [4, 5].

Thus, the newly redefined criteria allow for three possible research diagnoses: (1) asymptomatic neurocognitive impairment (ANI); (2) HIV-associated mild neurocognitive disorder (MND); and (3) HAD. However, if there are currently less cases of HAD as survival with chronic HIV-1 infection improves, the number of people harboring the virus in their CNS increases, leading to new HIV-1-related neurological manifestations. The prevalence of HAND therefore continues to rise, and less fulminant forms of HAND have become more common than their more severe predecessors [4, 5]. Incident cases of HAND are accelerating fastest among drug users, ethnic minorities, and women [6–9]. Moreover, it is becoming clear that the brain is an important reservoir for the virus, and that neurodegenerative and neuroinflammatory changes may continue despite HAART [8].

The principal manifestations of CNS in HIV infection result from neuronal injury and loss and from extensive damage to the dendritic and synaptic structures in the absence of neuronal loss. Neurons themselves are rarely infected by HIV-1, and neuronal damage is felt to be mainly indirect. In fact, the pathogenesis of HAND largely reflects the neurotoxicity of HIV-1 proteins [10].

HIV-1 infects resident microglia, periventricular macrophages, and some astrocytes [11], leading to increased production of cytokines and to release of HIV-1 proteins, the most likely neurotoxins, among which are the envelope (Env) proteins gp120 and gp41 and the nonstructural proteins Nef, Rev, Vpr, and Tat [9, 12, 13]. Soluble gp120 can induce apoptosis in a wide variety of cells including lymphocytes, cardiomyocytes, and neurons [14, 15]. HIV-1 gp120 may be directly neurotoxic at high concentrations [16]. Gp120-induced apoptosis has been demonstrated in studies in cortical cell cultures, in rat hippocampal slices, and by intracerebral injections in vivo [16]. Gp120 binds neuron cell membrane co-receptors (CCR3, CCR5, and CXCR4) and elicits apoptosis, apparently via G-protein-coupled pathways [17]. The HIV-1 *trans*-acting protein Tat, an essential protein for viral replication, is a key mediator of neurotoxicity. Tat is internalized by neurons primarily through lipoprotein-related protein receptor (LRP), by activation of NMDA receptor [18], and by interaction. It also interacts with integrins, VEGF receptor in endothelial cells, and possibly CXCR4 [19].

Tat can directly depolarize neuron membranes, independently of Na^+ flux [20] and may potentiate glutamate- and NMDA-triggered calcium fluxes and neurotoxicity [20]. It promotes excitotoxic neuron apoptosis [21, 22] by activating endoplasmic reticulum pathways to release intracellular calcium ($[Ca^{2+}]i$). Consequent dysregulation of calcium homeostasis [21, 23, 24] leads to mitochondrial calcium uptake, caspase activation, and, finally, neuronal death. Tat also increases levels of lipid peroxidation [22] by generating the reactive oxygen species (ROS) superoxide (O_2^-) and hydrogen peroxide (H_2O_2).

There are no perfect models for HAND. Several animal systems have been used to study the pathogenesis of HAND. Many of them are based on other lentiviruses [25–27]. However, only small percentages of animals develop neurological manifestations in these models and the costs for using these species may be high. Transgenic expression of gp120 in mice has been studied [28], but the gp120 in that model is mainly expressed in astrocytes, whereas in humans HIV-1 chiefly infects microglial cells. Some models of ongoing exposure to Tat have been developed. For example, GFAP-driven, doxycycline-inducible Tat transgenic mice have been useful for mechanistic studies of Tat contribution to HAND. However, the reported data concerning neuronal TUNEL positivity are still debated [29]. We [30–33] and others [34] have used model systems in which recombinant gp120, or Tat, proteins are directly injected into the striatum, in order to induce neuronal apoptosis. The neurotoxicity of such recombinant proteins is highly reproducible and can be used as an interesting tool for testing novel therapeutic interventions. Administration of recombinant proteins is useful in understanding the effects of HIV-1 gene products, and so their individual contribution to the pathogenesis of HAND, not only in term of apoptosis but also the deleterious effects related to gp120 or Tat [35–40]. However, HIV-1 infection of the brain is a chronic process, and its study would benefit from a model system allowing longer term exposure to HIV-1 gene product. This is in part the reason why we developed experimental models of chronic HIV-1 neurotoxicity based on recombinant SV40 (rSV40) vector-modified expression of gp120 [41] or Tat [42, 43] in the brain.

1.2 Apoptosis

Apoptosis is a process of programmed cell death (PCD) regulated by several genes and characterized by cell changes such as cell shrinkage, condensation of the cytoplasm with organelles appearing tightly packed, chromatin condensation, nuclear envelope discontinuous, DNA and nuclear fragmentation, and blebbing. Unlike necrosis, apoptosis produces cell fragments called apoptotic bodies that phagocytic cells are able to engulf and quickly remove before the contents of the cell can cause damage by spilling out onto surrounding cells. Enzymes are activated in a coordinated manner by cells destined to die, leading to an efficient degradation of their DNA, nuclear and cytoskeletal proteins, and all cellular components [44].

Apoptosis occurs in both physiological and pathological situations, in order to remove cells which are no longer useful or have been damaged. For example, the separation of fingers and toes in a developing human embryo occurs because cells of tissues located between the digits undergo apoptosis. Apoptosis is closely related to the immune system. It is related to the removal of cells recognized as nonself by cytotoxic T lymphocytes, as well as to the elimination of the same T lymphocytes, as well as B lymphocytes at the end of an immune response to avoid a harmful reaction to the body. Naturally occurring neuronal apoptosis peaks in the first postnatal week in rats and is negligible by postnatal day 21 [44]. Apoptosis is triggered through three signaling pathways: intrinsic and extrinsic (both caspase dependent), and the caspase-independent pathway.

The caspases family of proteases is conserved from nematodes through mammals. They are central to apoptotic death and are expressed as inactive zymogens that become cleaved during apoptosis [45]. Initiator caspases autoactivate and self-process upon recruitment to adaptor proteins. Then, they proceed to cleave and thereby activate the executioner/effector caspases. Activated executioner/effector caspases proceed to process key structural and nuclear proteins and thereby cause the disassembly and death of the cell [46]. Two major caspases pathways have been described: the intrinsic pathway is initiated by cytochrome c release from the mitochondrion while the extrinsic pathway is initiated by the binding of ligands to plasma-membrane death receptors [47].

Intrinsic apoptosis pathway is required for fetal and postnatal brain development, but is downregulated through the suppression of the expression of one of its key mediator, caspase-3 [46]. During stroke and neurodegenerative diseases, some caspases are upregulated in the brain [45]. Cerebral ischemia triggers both the intrinsic and extrinsic pathways of apoptosis [47, 48]. Mounting evidence suggests the involvement of caspases in the disease process associated with neurodegenerative diseases such as Alzheimer's disease (AD) [49] and amyotrophic lateral sclerosis (ALS) [46]. Caspase activation has also been documented in the brains of patients with HIV-1-associated dementia [12, 50]. The cysteine protease family of caspases is activated in immature neurons and contributes to apoptotic death [51]. Thus, a link between oxidative stress and activation of some caspases seems highly probable.

Apoptosis can be studied by transmission electron microscopy (TEM), as attested by numerous data in the literature. However, immunohistochemical methods, through the use of specific markers, provide ample information concerning the different patterns of activation during the apoptotic process [52].

Necrosis is a form of traumatic cell death that results from acute cellular injury, in contrast to apoptosis which is a highly regulated and controlled process that confers advantages during an organism's lifecycle [51]. In acute neuronal injury in the adult rat brain, irreversibly damaged neurons are morphologically necrotic and undergo a caspase-independent programmed cell death. This caspase-independent programmed pathway is associated with activation of the Ca^{2+}-dependent cysteine protease calpain. There are intracellular enzymes that also contribute to necrotic neuronal death [51, 52].

We examine here the different methods used for assessing apoptosis, neuronal loss, and expression of caspases in acute and chronic models of HAND.

2   Equipment, Materials, and Setup

2.1   Differentiation of COS-7 Cell Line into NT2-Neurons

- Cell culture hood
- COS-7 cell line [American Type Culture Collection (ATCC)]
- Dulbecco's modified eagle's medium (DMEM)
- 10 % calf serum (Hyclone, Logan, UT, USA), 2 mM L-glutamine and containing 1.5 g/l sodium bicarbonate, 4.5 g/l glucose, 1.0 mM sodium pyruvate, penicillin (200 U/ml), and streptomycin (100 µg/ml)
- Human N-tera2/cloneD1 (NT-2) [Stratagene (La Jolla, CA, USA)] induced to differentiate into NT2-Neurons (NT2-N)
- (DMEM/F-12) supplemented with glutamine and 10 % (v/v) calf serum
- Retinoic acid (Sigma Chemicals, MO, USA)
- Cytosine b-D-arabinofuranoside (ara C) (1 mM), uridine (1-b-D-ribofuranosyluracil) (Urd) (10 mM), and 5-fluoro-20-deoxyuridine (FUDR)
- Poly-D-lysine coated 24-well plates (Sigma Chemicals) and matrigel coated (BD Sciences, Bedford, MA) 4-chamber slides

2.2   Cell Culture of Primary Human Fetal Neuronal Cells

- Cell culture hood
- Hanks Balanced Salt Solution (HBSS) Ca^{2+} and Mg^{2+} free containing 0.05 % Trypsin and 100 U of DNAse
- Poly-D-lysine coated 24-well plates or 4-chamber slides
- DMEM/F-12 (Invitrogen, Carlsbad, CA, USA); cytosine arabinoside (ara C—1 mM)

2.3 ***Vector*** ***Production***	• Cell culture hood
	• COS7 cells
	• SV40 derived vectors
	• pT7[RSVLTR], in which transgene expression is controlled by the Rous Sarcoma Virus long terminal repeat (RSV-LTR) as a promoter
	• SOD1 and GPx1 transgenes; gp120 and Tat as transgenes
	• PCR

2.3 ***Vector Production***

- Cell culture hood
- COS7 cells
- SV40 derived vectors
- pT7[RSVLTR], in which transgene expression is controlled by the Rous Sarcoma Virus long terminal repeat (RSV-LTR) as a promoter
- SOD1 and GPx1 transgenes; gp120 and Tat as transgenes
- PCR

2.4 ***Stereotaxic Injection of gp12, Tat, and Vectors***

- Stereotaxic apparatus (Stoelting Corp., Wood Dale, IL)
- Feedback-controlled heater (Harvard Apparatus, Boston, MA)
- Glass micropipettes puller (World Precisions Instruments, Inc., Sarasota, FL)
- Picospritzer II (General Valve Corp., Fairfield, NJ)

2.5 ***Microscopy***

- Transmitted light microscope
- Fluorescence microscope equipped with appropriate filter sets
- Personal computer
- Image-Pro Plus software (MediaCybernetics, Bethesda, MD)
- Adobe Photoshop 6.0 software (Adobe Systems, San Jose, CA)
- Transmission electron microscope

2.6 ***General Equipment and Reagents for Tissue Preparation***

- Phosphate-buffered saline (PBS) 0.1 M pH 7.4
- Fixatives. *see* **Note 1**
- Slide oven
- OCT medium
- General histology slides and coverslips. *see* **Note 2**
- Cryostat
- Ultramicrotome

2.7 ***TUNEL Assay***

- TdT-mediated dUTP nick end labeling kit (Roche Diagnostics, Indianapolis, IN)
- Permeabilization solution: 0.1 % Triton X-100, 0.1 sodium citrate in PBS

2.8 ***Immuno-fluorescence (IMF)***

- 10 % normal goat serum (NGS) or 10 % normal donkey serum (according to host species of primary antibodies)
- Primary antibodies. *see* Tables 1 and 2
- Fluorochrome-conjugated secondary antibodies. *see* Tables 1 and 2. *see* **Note 3**
- Vectashield mounting medium hard set with 4′,6-diamidino-2-phenylindole (DAPI) (Vector Laboratories, Burlingame, CA)

Table 1
Primary and secondary antibodies

Primary antibodies			
Type of marker	Antigen	Type of antibody	Suggested supplier
Apoptosis	Active caspase 3	Rabbit polyclonal	Santa Cruz, Santa Cruz, CA
	Active caspase 6	Goat polyclonal	Santa Cruz, Santa Cruz, CA
	Active caspase 8	Rabbit polyclonal	Santa Cruz, Santa Cruz, CA
	Active caspase 9	Mouse monoclonal	Santa Cruz, Santa Cruz, CA
Cell type	NeuN	Mouse monoclonal	Chemicon International, Temecula, CA
	MAP-2	Mouse monoclonal	InVitrogen, Carlsbad, CA
	DAT	Rabbit polyclonal	Santa Cruz, Santa Cruz, CA
	TH	Mouse monoclonal	Immunostar, Hudson, WI
	CD68	Mouse monoclonal	Serotec, Oxford, UK
	Iba-1	Rabbit polyclonal	Waco Chemicals, Osaka, Japan
	CD11b	Mouse monoclonal	Accurate Chemicals, Westbury, NY
Other markers	SOD1	Rabbit polyclonal	Stressgen, Victoria, BC, Canada
Secondary antibodies			
Host species	Specificity	Fluorescence	Suggested supplier
Goat	Mouse	FITC	Sigma, Saint-Louis, MO
	Mouse	TRITC	Sigma, Saint-Louis, MO
	Rabbit	TRITC	Sigma, Saint-Louis, MO
	Sheep	FITC	Sigma, Saint-Louis, MO
Rabbit	Goat	Cy3	Sigma, Saint-Louis, MO
Donkey	Mouse	FITC	Jackson Immunoresearch Laboratories, West Grove, PA
	Mouse	TRITC	Jackson Immunoresearch Laboratories, West Grove, PA
	Rabbit	Cy3	Jackson Immunoresearch Laboratories, West Grove, PA
	Goat	Cy3	Jackson Immunoresearch Laboratories, West Grove, PA

2.8.1 IMF on Cell Cultures

- Triton X-100
- Paraformaldehyde
- PBS; Normal goat serum

2.8.2 IMF on Cryostat Sections

- PBS; Normal goat serum
- Cryoprotecting solution (30 % sucrose in PBS)
- OCT medium
- Cryostat

2.9 Morphometry

- Personal computer
- Image-Pro Plus software (MediaCybernetics, Bethesda, MD)
- Adobe Photoshop 6.0 software (Adobe Systems, San Jose, CA)

Table 2
Immunofluorescence parameters

Type of marker	Primary antibody	Dilution–incubation	Secondary antibody	Dilution–incubation
Apoptosis	Active caspase 3	1:100 (60 min)	TRITC goat anti-rabbit FITC sheep anti-rabbit Cy3 donkey anti-rabbit	1:100 (60 min)
	Active caspase 6	1:100 (60 min)	Cy3 rabbit anti-goat Cy3 donkey anti-goat	1:100 (60 min)
	Active caspase 8	1:100 (60 min)	TRITC goat anti-rabbit FITC sheep anti-rabbit Cy3 donkey anti-rabbit	1:100 (60 min)
	Active caspase 9	1:100 (60 min)	FITC and TRITC goat anti-mouse FITC and TRITC donkey anti-mouse	1:100 (60 min)
Cell type	NeuN	1:100 (60 min)	FITC and TRITC goat anti-mouse FITC and TRITC donkey anti-mouse	1:100 (60 min)
	MAP-2	1:100 (60 min)	FITC and TRITC goat anti-mouse	1:100 (60 min)
	DAT	1:50 (60 min)	TRITC goat anti-rabbit FITC sheep anti-rabbit	1:100 (60 min)
	TH	1:100 (60 min)	FITC and TRITC goat anti-mouse FITC and TRITC goat anti-mouse	1:100 (60 min)
	CD68	1:100 (60 min)	FITC and TRITC goat anti-mouse	1:100 (60 min)
	Iba-1	1:100 (60 min)	TRITC goat anti-rabbit FITC sheep anti-rabbit	1:100 (60 min)
	CD11b	1:100 (60 min)	FITC and TRITC goat anti-mouse	1:100 (60 min)
Other markers	SOD1	1:100 (60 min)	TRITC goat anti-rabbit FITC sheep anti-rabbit	1:100 (60 min)

2.10 Neurotrace

- Neurotrace (NT; Molecular Probes, Inc., Eugene, OR)

3 Procedures

3.1 Differentiation of COS-7 Cell Line into NT2-Neurons and Exposure to HIV-1 Neurotoxin

COS-7 cell line was obtained from American Type Culture Collection (ATCC) and was maintained in Dulbecco's modified eagle's medium (DMEM) supplemented with 10 % calf serum (Hyclone, Logan, UT, USA), 2 mM L-glutamine and containing 1.5 g/l sodium bicarbonate, 4.5 g/l glucose, 1.0 mM sodium pyruvate, penicillin (200 U/ml), and streptomycin (100 μg/ml). Human N-tera2/cloneD1 (NT-2), derived from a teratocarcinoma was obtained from Stratagene (La Jolla, CA, USA) and induced to differentiate into NT2-Neurons (NT2-N) according to manufacturer's instructions. The cells were propagated in DMEM with nutrient mixture F-12 (DMEM/F-12) supplemented with glutamine and 10 % (v/v) calf serum. Differentiation to neurons was induced by adding 10 mM retinoic acid (Sigma Chemicals, MO, USA). After the first replating, the cells were treated with mitotic inhibitors: cytosine b-D-arabinofuranoside (ara C) (1 mM), uridine (1-b-D-ribofuranosyluracil) (Urd) (10 mM), and 5-fluoro-20-deoxyuridine (FUDR) (10 mM) for 3 weeks. After further enrichment, and selective trypsinization, highly enriched neurons were harvested 2–3 weeks later. The neuron cultures obtained were >95 % pure neurons and were placed on poly-D-lysine (Sigma Chemicals) and matrigel coated (BD Sciences, Bedford, MA) 4-chamber slides or 24-well tissue culture plates. The neurons were characterized by immunocytochemistry using MAP-2 and NeuN antibodies (Chemicon International Inc., Temecula, CA). Apoptosis was induced in NT2-neurons by incubation of the cells during 2 days with different concentrations of recombinant HIV-1-gp120-Ba-L (0.1–100 ng/ml). Cells were washed and cultured for another 3 days before TUNEL assay. *see* **Note 4**.

3.2 Cell Cultures of Primary Human Fetal Neurons and Induction of Apoptosis

Fetal brain (obtained from the Human Fetal Tissue Bank, Albert Einstein College of Medicine, Bronx, NY, USA) was homogenized in Hanks Balanced Salt Solution (HBSS) Ca^{2+} and Mg^{2+} free containing 0.05 % Trypsin and 100 U of DNAse. Mixed brain cultures were passed through 170 mm Nylon mesh, then plated in poly-D-lysine coated 24-well plates or 4-chamber slides. Nonadherent cells were removed by washing with DMEM/F-12 (Invitrogen, Carlsbad, CA, USA). Adherent neuronal cultures were treated with cytosine arabinoside (ara C—1 mM) for 2 weeks. Human primary neurons were exposed to gp120 according to the same protocol used for NT2-neurons. The human primary neurons were subsequently immunostained with MAP-2 and NeuN antibodies.

3.3 Vector Production

The general principles for making recombinant, *Tag*-deleted, replication-defective SV40 viral vectors have been previously reported [53]. SOD1 and GPx1 transgenes were subcloned into pT7[RSVLTR], in which transgene expression is controlled by the Rous Sarcoma Virus long terminal repeat (RSV-LTR) as a promoter. Tat and gp120 expression in SV(Tat) and SV(gp120) respectively is driven by RSV-LTR [41–43]. The cloned rSV40 genome was excised from its carrier plasmid, gel-purified, and recircularized, then transfected into COS-7 cells. These cells supply *in trans* large T-antigen (Tag) and SV40 capsid proteins, which are needed to produce recombinant replication-defective SV40 viral vectors. Crude virus stocks were prepared as cell lysates, then band-purified by discontinuous sucrose density gradient ultracentrifugation and titered by Q-PCR. SV(BUGT), which was used here as negative control vector, has been reported. We reported successful transgene expression using rSV40-derived vectors in various animal systems using different protocols [54–56].

3.4 Experimental Design for In Vivo Experiments

Protocols for injecting and euthanizing animals were approved by the Thomas Jefferson University (Philadelphia, PA, USA) IACUC and are consistent with AAALAC standards. To test gp120-induced apoptosis and neuronal loss, different doses of gp120 (100, 250, 500 ng) were injected unilaterally into the CP of the rat brain stereotaxically. For the different doses considered and for all following experiments of the study, gp120 was injected in 1 μl saline. Recombinant HIV-1 BaL gp120 was obtained through the NIH AIDS Research & Reference Reagent Program, Division of AIDS, NIAID, NIH, Germantown, MD. Saline was used as the negative control, as was the contralateral side of the unilaterally injected brains. Brains were harvested at 6 h and on 1, 2, 4, 7, and 14 days after the injection and were processed for histological staining (H&E); TUNEL assay for apoptosis; and immunocytochemistry of markers of apoptosis (caspases 3, 6, 8, 9). Neuronal loss was assessed by using histological staining of Neurotrace (NT), a marker of neurons, and immunocytochemistry of neuN, a marker of mature neuronal neurons, and DAT, the dopamine transporter. The SN of the brains harvested 1, 4, and 7 days after injection of 500 ng gp120 into the CP was immunostained for TH, as were SN sections of brains 7 days after injection with the three different gp120 doses. SN sections were also tested by TUNEL assay (Roche, Indianapolis, IN) for assessing the presence of putative apoptotic cells in the SN after injection of 500 ng gp120 into the CP. Five rats were used for each dose of gp120 for a particular time point, and four control rats were injected with saline for each time point. see **Note 5**.

3.5 Stereotaxic Injection of gp12, Tat, and Vectors

Rats were anesthetized with isoflurane UPS (Baxter Healthcare Corp., Deerfield, IL) (1.0 U isoflurane/1.5 l O_2 per min) and placed in a stereotaxic apparatus (Stoelting Corp., Wood Dale, IL) for cranial surgery. Body temperature was maintained at 37 °C by using a feedback-controlled heater (Harvard Apparatus, Boston, MA). Glass micropipettes (1.2 mm outer diameter; World Precisions Instruments, Inc., Sarasota, FL) with tip diameters of 15 μm were backfilled with 1 μl saline containing gp120. The gp120-filled micropipettes were placed in the CP using coordinates obtained from the rat brain atlas of Paxinos and Watson [57]. For injection into the CP, a burr hole was placed +0.48 mm anterior to bregma and −3.0 mm lateral to the sagittal suture. Once centered, the micropipette was placed 6.0 mm ventral from the top of the brain. Injection of Tat was performed in a similar way. The vector was given by a Picospritzer II (General Valve Corp., Fairfield, NJ) pulse of compressed N2 duration 10 ms at 20 psi until the fluid was completely ejected from the pipette. Following surgery, animals were housed individually with free access to water and food [31, 32]. *see* **Note 5**.

3.6 Preparation of Sections

After a variable survival period, rats were anesthetized via intraperitoneal injection of sodium pentobarbital (Abbott Laboratories, North Chicago, IL) at 60 mg/kg and perfused transcardially though the ascending aorta with 10 ml heparinized saline followed by 500 ml ice-cold 4 % paraformaldehyde (Electron Microscopy Sciences, Fort Washington, PA) in 0.1 M phosphate buffer (pH 7.4). Immediately following perfusion fixation, the rat brains were dissected out, placed in 4 % paraformaldehyde for 24 h, then in a 30 % sucrose solution for 24 h, and finally frozen in methyl butane cooled in liquid nitrogen. The brains were cut transversally on a cryostat (10 μm sections). The area of the SN pars compacta was determined using the atlas of Paxinos and Watson [57]. All sections cut in this area were harvested on slides. Some of these slides were tested for TH immunoreactivity to confirm that the sections were part of the SN pars compacta [31, 32].

3.7 TUNEL Assay

TUNEL (terminal deoxynucleotidyl transferase dUTP nick end labeling) assay was performed using a commercial kit and according to the protocol recommended by the manufacturer (Roche Diagnostics, Indianapolis, IN). TUNEL assays are a valuable method for detecting DNA fragmentation, which is a hallmark of apoptosis, or programmed cell death. In TUNEL assay, an enzyme known as terminal deoxynucleotidyl transferase (TdT) identifies nicks, or points of fragmentation, in the sample DNA. TdT catalyzes the addition of dUTP nucleotides that have been labeled previously, for subsequent detection. Permeabilization of cells or tissues with 0.1 % Triton X-100 was the first step of the procedure. *see* **Notes 6 and 7**.

3.8 Immunocytochemistry/ Histochemistry

3.8.1 Immunocytochemistry of Cell Cultures

Cells were grown on either 24-well plates or four-chamber slides treated with poly-D-lysine and matrigel. Cells were fixed with 1 % paraformaldehyde for 30 min on ice and permeabilized with 0.1 % Triton X-100 in sodium citrate buffer. Nonspecific binding was blocked by treating cells with normal serum from the animal species in which the secondary antibody was raised, then immunostained with either anti-MAP-2 (1:100) or anti-Neu-N(1:100) (Molecular Probes Invitrogen, Carlsbad, CA, USA) for 1 h on ice. After extensive washes in PBS containing 1 % BSA, secondaries antibodies conjugated with Alexa Fluor 488 (Molecular Probes, Oregon, CA, USA) or TRITC (Sigma Chemicals) were added. The cells were washed and analyzed on fluorescence microscope [30, 39].

3.8.2 Immunohisto-chemistry on Cryostat Sections

Coronal cryostat sections (10 μm thick) are cut after cryoprotection and processed for indirect immunohistochemistry. Sections are first incubated for 60 min with 10 % normal goat serum, or 10 % normal donkey serum in PBS to block nonspecific binding. They are then incubated with antibodies diluted according to manufacturers' recommendations: 1 h with primary antibody, then 1 h with secondary antibody diluted 1:100. Double histochemistry is performed according to standard protocols [55]. Mounting media contain DAPI to stain nuclei. Negative controls are performed each time immunostaining was done and consisted of preincubation with PBS, substitution of non-immune isotype-matched control antibodies for the primary antibody, and/or omission of the primary antibody. Further details can be found in [54, 55, 58].

3.9 Staining of Neurons Using NeuroTrace

After rehydration in 0.1 M PBS, pH 7.2, sections were treated with PBS plus 0.1 % Triton X-100 10 min, washed twice for 5 min in PBS, then stained by NeuroTrace (NT) (Molecular Probes, Inc., Eugene, OR) (1:100), a fluorescent Nissl stain, for 20 min at room temperature. Sections were washed in PBS plus 0.1 % Triton X-100, then twice with PBS, then let stand for 2 h at room temperature in PBS before being counterstained with DAPI. Combination NT + antibody staining was performed using primary and secondary antibodies staining first (see above), followed by staining with the NT fluorescent Nissl stain. For antibody, TUNEL, and NT staining, immunohistochemistry was the first step, followed by TUNEL assay, then by NT staining. All experiments were repeated three times and test and control slides were stained the same day. Experiments were repeated three times and were done the same day for the different sections considered [54, 55].

3.10 General Morphology

Microscopic morphology of the brain was assessed by neutral red (NR), and H&E staining of cryostat sections [35].

3.11 Transmission Electron Microscopy (TEM)

Adult Sprague–Dawley rats were used for this study, with or without injection of gp120 in the CP. One day after injection of gp120, the rats were deeply anesthetized with sodium pentobarbital (60 mg/kg) and perfused transcardially through the ascending aorta with (1) 10 ml heparinized saline, (2) 50 ml of 3.75 % acrolein (Electron Microscopy Sciences, Fort Washington, PA, USA), and 200 ml of 2 % formaldehyde in 0.1 M PB, pH 7.4. Immediately after perfusion fixation, brains were removed, sectioned coronally, and postfixed in the same fixative overnight at 4 °C. Alternate 40-μm thick sections through the CP were processed for electron microscopy. Sections containing the CP were placed for 30 min in 1 % sodium borohydride in 0.1 M PB and collected into 0.1 M PB to remove reactive aldehydes. Then sections were rinsed extensively in 0.1 M PB. Subsequently, sections were rinsed three times with 0.1 M TBS, followed by rinses with 0.1 M PB and 0.01 M phosphate-buffered saline (PBS; pH 7.4). Sections were then incubated in 2 % glutaraldehyde (Electron Microscopy Sciences) in 0.01 M PBS for 10 min followed by washes in 0.01 M PBS and 0.2 M sodium citrate buffer (pH 7.4). Following washes, tissues were rinsed in 0.2 M citrate buffer and 0.1 M PB, and incubated in 2 % osmium tetroxide (Electron Microscopy Sciences) in 0.1 M PB for 1 h, washed in 0.1 M PB, dehydrated in an ascending series of ethanol followed by propylene oxide and flat embedded in Epon 812 (Electron Microscopy Sciences). Thin sections of approximately 50–100 nm in thickness were cut with a diamond knife (Diatome-US, Fort Washington, PA, USA) using a Leica Ultracut (Leica Microsystems, Wetzlar, Germany). Sections were collected on copper mesh grids, examined with an electron microscope (Morgagni, Fei Company, Hillsboro, OR, USA), and digital images were captured using the AMT advantage HR/HR-B CCD camera system (Advance Microscopy Techniques Corp., Danvers, MA, USA). Figures were assembled and adjusted for brightness and contrast in Adobe Photoshop [32].

3.12 Morphometry

3.12.1 Morphometry of Apoptotic Cells

Caspases- and TUNEL-positive cells were enumerated manually on the injected and uninjected sides in the whole CP of animals injected with gp120, or saline, in at least five consecutive sections using a computerized imaging system (Image-Pro Plus, MediaCybernetics, Bethesda, MD). In all cases, the final number was an average of results measured in the different sections. This procedure allows quantitative and relative comparisons among different time point. The results were expressed as percentages of NT-positive cells.

There were no TUNEL-positive cells when the CP was injected with saline (negative control). For a positive control, we injected intraperitoneally (IP) kainic acid, a drug known for inducing apoptosis in hippocampus, and assayed neurons of the dentate gyrus (DG). At all doses of gp120, TUNEL-positive cells peaked 1 day after the injection, and apoptotic nuclei were confirmed by transmission electron microscopy [31, 32].

3.12.2 Morphometry of Neuronal Loss

A computerized imaging system (Image-Pro Plus, MediaCybernetics, Bethesda, MD) was used to quantify the area of striatal tissue loss. The area of tissue loss was determined using NR-stained sections using the imaging system. Computer-assisted tracing of the perimeter of the striatal tissue loss surrounding the injection site as well as the whole CP was conducted to determine area measures. A ratio of the area of tissue loss compared to the whole CP area was determined for each section that was considered. A total of 20 sections (one section every 300 μm; ten sections rostral and ten sections caudal to the injection site) per animal were used. This procedure was similar to the one previously described [34]. The length of the extent of tissue damage was determined on serial sections of the whole brain stained by NR [35].

We assessed the volume of the lesion on NR-stained sections according to the Cavalieri estimate of volume by overlaying a grid of known spacing over the image of the section on the monitor and by counting the number of points which fall upon the region of interest, and finally by calculating the overall volume. The region of interest corresponded to Plate 10 through Plate 31 of the Paxinos and Watson rat brain atlas [57]. Approximately 12 sections per brain were analyzed. Using this method, our estimation of the CP volume in control uninjected rats was 56.1 ± 4.8 mm^3, which is close to previous published value of 50.1 ± 3.7 mm^3. A ratio of the volume of tissue loss compared to the whole CP volume was also determined and expressed as percentage [35].

Neuronal loss in the CP was assessed by identifying NT-positive cells and DAT-positive structures on the whole CP in a section and by counting them using the Image-Pro Plus computerized imaging system. A total of 20 sections (one section every 100 μm; ten sections rostral and ten sections caudal to the injection site) per animal were used. A ratio of the number of NT-positive cells and DAT-positive structures on the injected side compared to the number of NT-positive cells and DAT-positive structures on the contralateral uninjected side was calculated [35].

TH-positive cells were enumerated on ten sections of the SN (30 μm apart) on the ipsilateral and contralateral side of the injection site, and a ratio of the number of TH-positive cells on the ipsilateral side compared to the number of TH-positive cells on the contralateral side was calculated [35].

4   Typical/Anticipated Results

The following figures show the methods that can be useful in the evaluation of apoptosis and neuronal loss in animal models of HAND.

4.1   Demonstration of Apoptosis Induced by gp120 in NT2-Neurons and Primary Neurons

Exposure of NT2-neurons and primary neurons to HIV-1 envelope glycoprotein gp120 resulted in apoptosis assessed by TUNEL assay. Apoptosis was induced in NT2-neurons by incubation of the cells during 2 days with different concentrations of recombinant HIV-1-gp120-Ba-L (0.1–100 ng/ml). Cells were washed and cultured for another 3 days before TUNEL assay. NT2-neurons were immunostained using antibodies against MAP-2 (Fig. 1a). There was a relationship between the concentration of gp120 and the number of apoptotic NT-neurons (Fig. 1a). Human primary neurons were

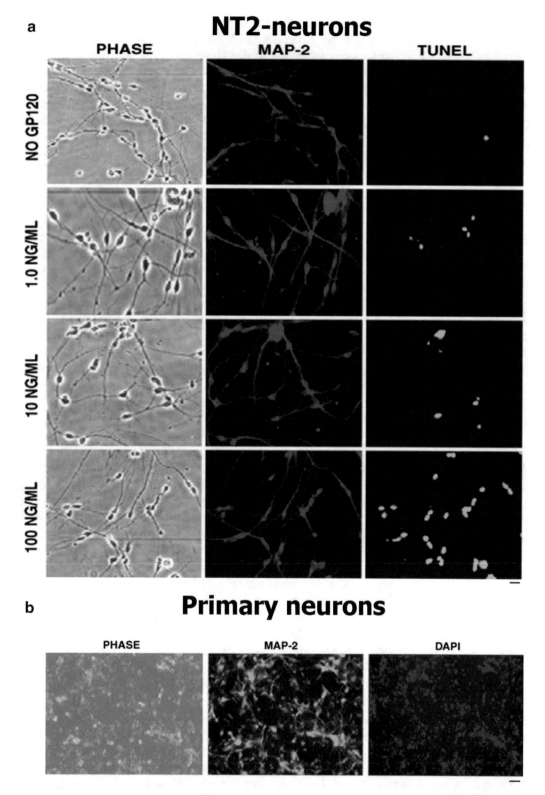

Fig. 1 Apoptosis of NT2-neurons and human primary neurons following exposure to HIV-1 envelope glycoprotein gp120. (**a**) NT2-neurons were incubated during 2 days with different concentrations of

immunostained for MAP-2 and were exposed to gp120 according to the same protocol used for NT2-neurons. Primary neurons underwent apoptosis, as described for NT2-neurons.

4.2 Evaluation of Apoptosis in Caudate-Putamen Injected with gp120

We used TUNEL assay and TEM to show apoptotic cells following gp120 injection into the CP. Stereotaxic injection of gp120 into the rat caudate-putamen (CP) (Fig. 2a) induced apoptosis evaluated by TUNEL assay (Fig. 2b) and TEM (Fig. 2d). Morphometric analysis shows a peak of apoptotic cells 24 h after intra-CP gp120 injection (Fig. 2c).

4.3 Gp120-Induced Apoptotic Cells Are Mostly Neurons

We performed immunohistochemistry to determine the nature of apoptotic cells. Figure 3 shows that TUNEL-positive cells seen after intra-CP injection of gp120 were mainly neurons (either Neurotrace-NT- or neuN-positive cells), and rarely macrophages (CD68-positive cells). The number of apoptotic cells was related to the concentration of gp120.

4.4 Expression of Caspases in Apoptotic Cells Following Intra-CP Injection of gp120

Caspases are expressed in apoptotic cells. We showed caspase expression by immunohistochemistry. Most of caspase-expressing cells were TUNEL positive following injection of gp120 into the CP (Fig. 4a). Caspases were mainly colocalizing with neuronal markers (NT or neuN) (Fig. 4b). Whatever the caspase considered, morphometric analysis showed that the number of caspase-positive cells peaked 24 h after injection of gp120 into the CP (Fig. 4c), with most caspase-positive cells being TUNEL positive (Fig. 4d).

4.5 Intra-CP Injection of gp120 Induces Local and at Distance Neuronal Loss

Besides apoptosis, neuronal loss was also observed and evaluated by morphometry following intra-CP gp120 injection. Figure 5a shows an area of the CP devoid of cells following injection of gp120 in the same structure. A loss of neurons (neuN-positive cells) was observed after injection of gp120 into the CP (Fig. 5b). This neuronal loss not only involves DAT-positive, CP neurons (Fig. 5c) but also includes dopaminergic neurons probably by retrograde degeneration (Fig. 5d, f). There was a relationship between the extent of loss of dopaminergic neurons and the concentrations of gp120 on one hand, and the time after intra-CP gp120 on the other hand. Similarly, there was a relationship between the volume of the lesion and the concentrations of gp120.

Fig. 1 (continued) recombinant HIV-1-gp120-Ba-L (0.1–100 ng/ml). Cells were washed and cultured for another 3 days before TUNEL assay. NT2-neurons were immunostained using antibodies against MAP-2. Apoptosis was assessed by TUNEL assay. There was a relationship between the concentration of gp120 and the number of TUNEL-positive NT-neurons. Bar: 40 μm. (**b**) Human primary neurons were exposed to gp120 according to the same protocol used for NT2-neurons. Primary neurons underwent apoptosis, as described for NT2-neurons. Human primary neurons were immunostained for MAP-2 and counterstained with DAPI. Bar: 40 μm. Modified from [30] with permission from Nature Publishing Group. MacMillan Publishers Limited

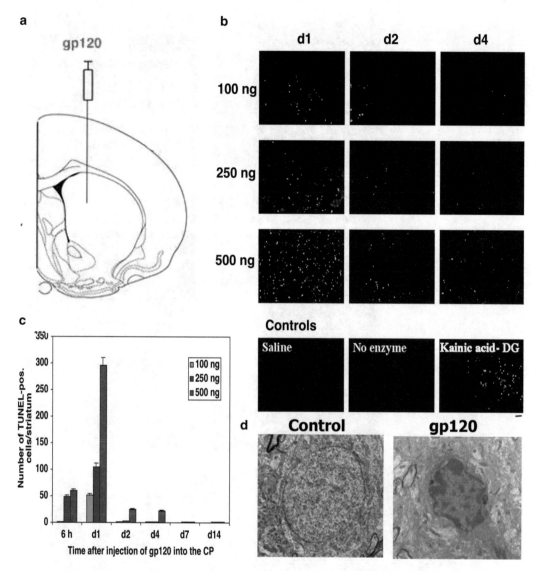

Fig. 2 Injection of gp120 into the caudate-putamen induces apoptosis. (**a**) HIV-1 envelope glycoprotein gp120 was injected stereotaxically into the rat caudate-putamen (CP). (**b**) TUNEL assay demonstrated numerous apoptotic cells following the injection of gp120. The number of TUNEL-positive cells was related to the concentration of gp120 injected into the CP. Negative controls included injection of saline instead of gp120, and section of brain injected with gp120 incubated with TUNEL assay, but without the enzyme TdT. Positive control for TUNEL assay consisted of a section of brain (focused on the dentate gyrus—DG—of the hippocampus) of rat injected with kainic acid, a molecule known to be excitotoxic. Bar: 80 μm. (**c**) Morphometric analysis showed a peak of apoptotic cells 24 h after intra-CP gp120 injection. (**d**) TEM demonstrated apoptotic nuclei. Modified from [32] with permission from Nature Publishing Group. MacMillan Publishers Limited

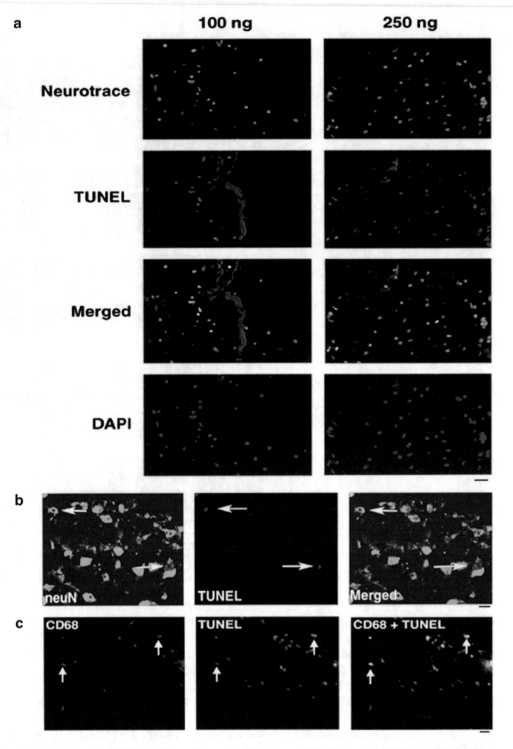

Fig. 3 Gp120-induced TUNEL-positive cells are mainly neurons. (**a**) Numerous TUNEL-positive cells were stained by Neurotrace (NT), a neuronal marker. Note that more numerous apoptotic cells were seen with 250 ng gp120 than with 100 ng gp120. Bar: 100 μm. (**b**) Similarly, apoptotic nuclei were identified in neurons immunostained for neuN, another neuronal marker. Bar: 80 μm. (**c**) TUNEL-positive cells were rarely macrophages (CD68-positive cells). The number of apoptotic cells was related to the concentration of gp120. Bar: 80 μm. Modified from [32] with permission from Nature Publishing Group. MacMillan Publishers Limited

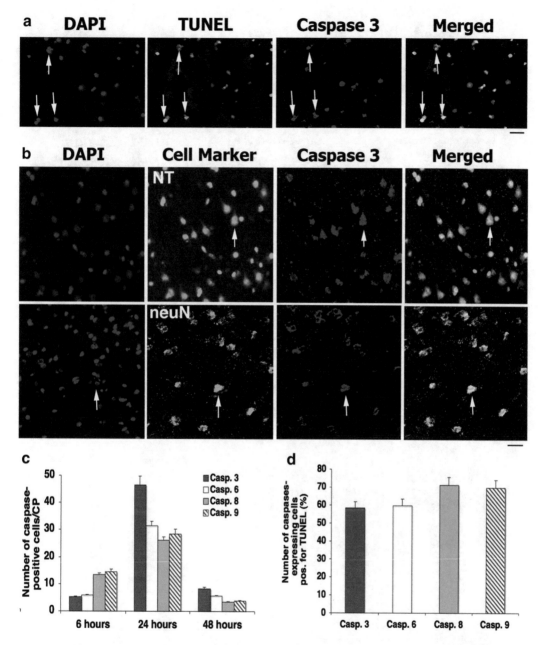

Fig. 4 Intra-CP injection of gp120 induces expression of caspases. (**a**) Injection of gp120 into the CP caused overexpression of caspases 3, 6, 8, and 9. Most caspase 3-expressing cells were apoptotic, TUNEL positive. Bar: 80 μm. (**b**) Caspase 3 was mainly expressed in neurons, stained for neuN or NT. Bar: 60 μm. (**c**) Whatever the nature of the caspase overexpressed, morphometric analysis showed that the number of caspase-positive cells peaked 24 h after injection of gp120 into the CP. (**d**) Most caspase-positive cells were TUNEL positive. Modified from [33] with permission from Elsevier

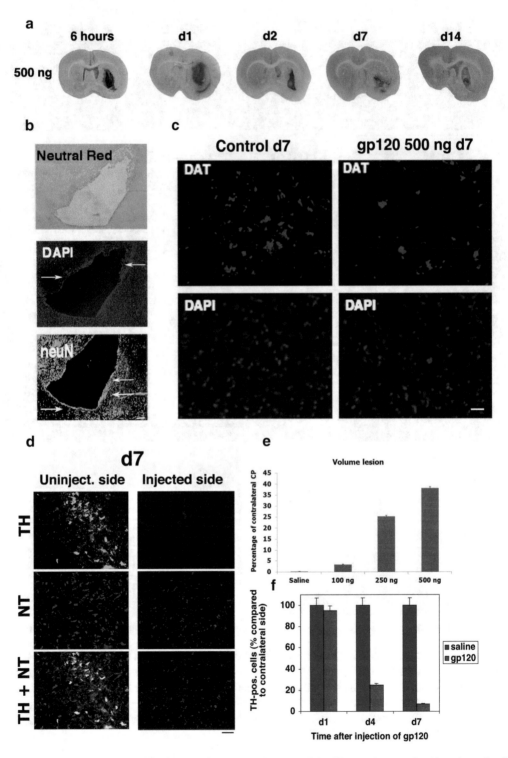

Fig. 5 Neuronal loss after intra-CP injection of gp120. (**a**) An area of the CP was characterized by a loss of cells following injection of gp120 in the same structure. (**b**) A loss of neurons (neuN-positive cells) was observed after injection of gp120 into the CP. Bar: 150 μm. (**c**) Reduction in number of DAT-positive cells in the CP after

4.6   Reduction of Oxidative Stress Limits gp120-Induced Apoptosis

Oxidative stress is involved in numerous neurodegenerative disorders, including HAND. We first demonstrated by immunofluorescence that gene delivery of antioxidant enzymes SOD1 and GPx1 in the CP resulted in strong expression of the transgene in neurons (Fig. 6a). Prior gene delivery of antioxidant enzymes to the CP by rSV40 vectors [respectively SV(SOD1) and SV (GPx1)] mitigated gp120-induced apoptosis (Fig. 6b, c). SV (BUGT) was used as a control vector. SV40-derived vectors were administered in the CP, and gp120 was injected at different times (either 4, 8, or 24 weeks) after inoculation of the vector. TUNEL assay was performed 1 day after injection of gp120 to assess the extent of apoptosis.

4.7   Tat Injection Causes Apoptosis

By contrast with the situation observed with gp120, apoptosis peaked 2 days after injection of Tat in the CP (Fig. 7a). Apoptotic cells were mostly neurons (neuN-positive cells), and more rarely macrophages (CD11b and Iba-1-positive cells) (Fig. 7b). This result shows that determining the time course of apoptosis is different between gp120- and Tat-induced insults.

4.8   SV40-Based Protracted Exposure to gp120 and Tat

Injection of gp120 and Tat in the CP provides interesting animal models for testing novel therapeutic approaches. However, the lesions are acute in these models. This is why we developed models where exposure to gp120 or Tat is protracted, by inserting the corresponding genes into rSV40 vectors. Injection of SV(gp120) and SV(tat) in the rat CP induced apoptosis (Fig. 8a, c), observed as long as 12 weeks (Fig. 8b). TUNEL-positive cells were mainly neurons, and more rarely macrophages (CD11b positive cells) (Fig. 8a–d).

5   Notes and Troubleshootings

1. The choice of fixative depends on the material to be studied: studies in experimental animal models are usually carried out on 4 % paraformaldehyde (PFA) fixed samples after perfusion. Fixation for in vitro studies can be done with different concentrations of PFA. The fixative has a profound influence on the antigenicity of the various molecular components and thus experiments must be planned carefully according to the type/ s of molecules under study.

Fig. 5 (continued) gp120 injection into the CP. Bar: 45 μm. (**d**) Loss of dopaminergic neurons in the Substantia Nigra was observed and probably explained by retrograde degeneration. Bar: 80 μm. (**e**) Relationship between the volume of the lesion and the concentrations of gp120. (**f**) Relationship between the extent of loss of dopaminergic neurons and the concentrations of gp120 on one hand (not shown), and the time after intra-CP injection of 500 ng gp120 on the other hand. Modified from [35] with permission from Elsevier

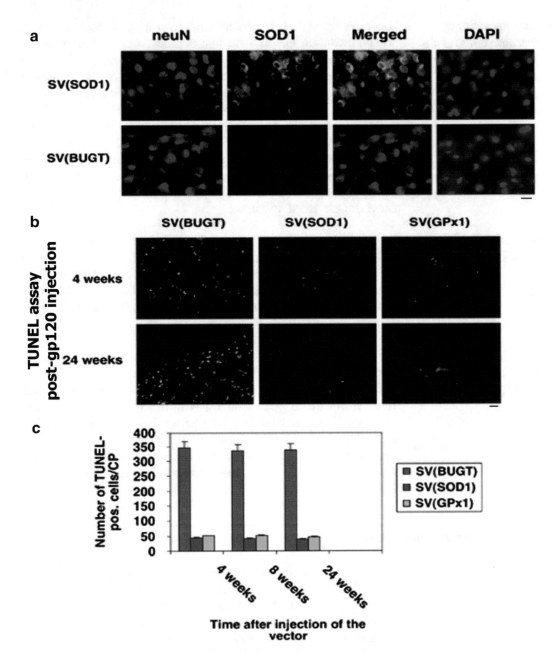

Fig. 6 Gene delivery of antioxidant enzymes mitigates gp120-induced apoptosis. (**a**) Gene delivery of antioxidant enzymes SOD1 and GPx1 in the CP results in strong expression of the transgene in neurons. (**b**) Prior gene delivery of antioxidant enzymes to the CP by rSV40 vector [SV(SOD1)] mitigates gp120-induced apoptosis. SV(BUGT) was used as a control vector. SV40-derived vectors were administered in the CP, and gp120 was injected at different times (either 4, 8, or 24 weeks) after inoculation of the vector. TUNEL assay was performed 1 day after injection of gp120. Bar: 30 μm. (**c**) Morphometric analysis showed that overexpression of antioxidant enzymes reduced the number of TUNEL-positive cells. Bar: 80 μm. Modified from [31, 32] with permission from Nature Publishing Group, MacMillan Publishers Limited

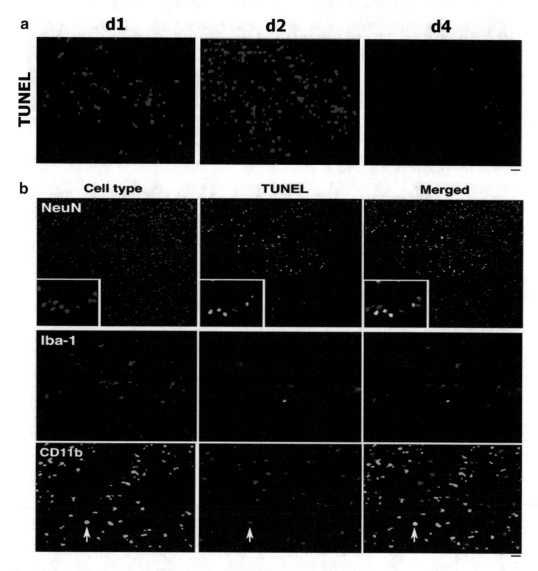

Fig. 7 Tat injection causes apoptosis. (**a**) As attested by TUNEL assay and confirmed by morphometry (not shown), apoptosis peaked 2 days after injection of Tat in the CP. Bar: 60 μm. (**b**) Apoptotic cells were mostly neurons (neuN-positive cells), and more rarely macrophages (CD11b and Iba-1-positive cells). Bar: 100 μm (*upper row*), 60 μm (*middle* and *lower rows*). Modified from [43] with permission from Elsevier

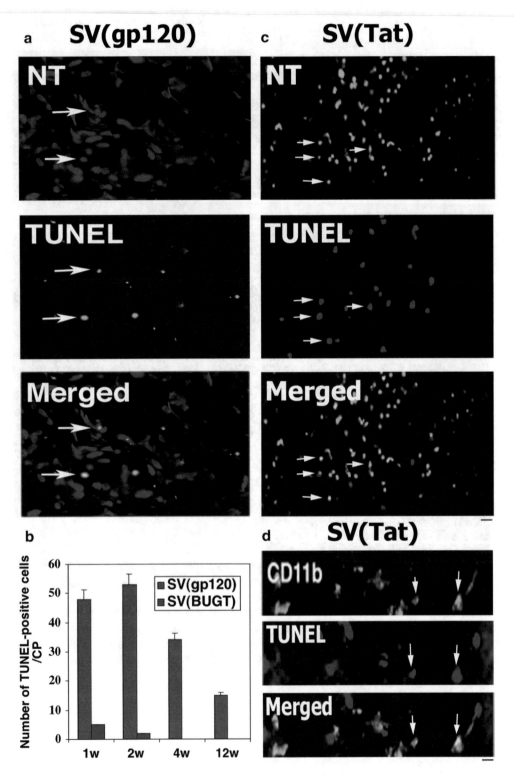

Fig. 8 SV40-based protracted exposure to gp120 and Tat. We developed models where exposure to gp120 or Tat is protracted, by inserting the corresponding genes into rSV40 vectors. (**a, c**) Injection of SV(gp120) and SV (tat) in the rat CP induced apoptosis. TUNEL-positive cells were mainly neurons. Bar: 60 μm in (**a**) and (**b**). (**b**) TUNEL-positive cells were observed as long as 12 weeks post-intra-CP injection. (**d**) Apoptotic cells were more rarely macrophages (CD11b positive cells). Bar: 30 μm. Modified from [41] and [43] with permission respectively from Elsevier and Lippincott Wolters Kluwer

2. Adhesion to slides can be improved by using commercially available slides with increased adhesiveness, e.g., Superfrost™ Plus and ColorFrost™ Plus Microscope Slides, Thermo Scientific Inc., Waltham, MA, or by immersing slides with poly-L-lysine, gelatin, or collagen.

3. Fluorochrome-conjugated secondary antibodies can be selected according to the specific needs of the study.

4. As a negative control for NT2-neurons and human primary neurons, immunostaining was performed with an isotype-matched immunoglobulin. Western analysis was also performed for MAP-2 and Neu-N using GAPDH as an internal loading control (not shown).

5. Anesthetics must be chosen according to the species used for experiments and the national and institutional regulations/guidelines for experimental animal care. Euthanasia must follow the guidelines issue by relevant veterinary authority, e.g., https://www.avma.org/kb/policies/documents/euthanasia.pdf

6. For TUNEL assay, the optimal temperature of incubation was 37 °C.

7. Positive control for TUNEL assay involved brains of rats injected with kainic acid (20 mg/kg, Sigma), a molecule known for inducing apoptosis in the hippocampus. Negative controls consisted of brains injected with saline instead of gp120, and brain injected with gp120, but incubated with the TUNEL solution without enzyme.

References

1. McArthur JC, Hoover DR, Bacellar H et al (1993) Dementia in AIDS patients: incidence and risk factors. Multicenter AIDS Cohort Study. Neurology 43:2245–2252

2. Major EO, Rausch D, Marra C et al (2000) HIV-associated dementia. Science 288:440–442

3. Koutsilieri E, Sopper S, Scheller C et al (2002) Parkinsonism in HIV dementia. J Neural Transm 109:767–775

4. Antinori A, Arendt G, Becker JT et al (2007) Updated research nosology for HIV-associated neurocognitive disorders. Neurology 69:1789–1799

5. Woods SP, Moore DJ, Weber E et al (2009) Cognitive neuropsychology of HIV-associated neurocognitive disorders. Neuropsychol Rev 19:152–168

6. McArthur JC, Brew BJ, Nath A (2005) Neurological complications of HIV infection. Lancet Neurol 4:543–555

7. Nath A, Sacktor N (2006) Influence of highly active antiretroviral therapy on persistence of HIV in the central nervous system. Curr Opin Neurol 19:358–361

8. Ances BM, Ellis RJ (2007) Dementia and neurocognitive disorders due to HIV-1 infection. Semin Neurol 27:86–92

9. Mattson MP, Haughey NJ, Nath A (2005) Cell death in HIV dementia. Cell Death Differ 12:893–904

10. Rumbaugh JA, Nath A (2006) Developments in HIV neuropathogenesis. Curr Pharm Des 12:1023–1044

11. Gonzalez-Scarano F, Martin-Garcia J (2005) The neuropathogenesis of AIDS. Nat Rev Immunol 5:69–81

12. Kaul M, Garden GA, Lipton SA (2001) Pathways to neuronal injury and apoptosis in HIV-associated dementia. Nature 410:988–994

13. van de Bovenkamp M, Nottet HS, Pereira CF (2002) Interactions of human immunodeficiency virus-1 proteins with neurons: possible role in the development of human immunodeficiency virus-1 associated dementia. Eur J Clin Invest 32:619–627

14. Garden GA, Guo W, Jayadev S et al (2004) HIV associated neurodegeneration requires p53 in neurons and microglia. FASEB J 18:1141–1143

15. Xu Y, Kulkosky J, Acheampong E et al (2004) HIV-1-mediated apoptosis of neuronal cells: proximal molecular mechanisms of HIV-1-induced encephalopathy. Proc Natl Acad Sci U S A 101:7070–7075

16. Meucci O, Fatatis A, Simen AA et al (1998) Chemokines regulate hippocampal neuronal signalling and gp120 neurotoxicity. Proc Natl Acad Sci U S A 95:14500–14505

17. Kaul M, Lipton SA (1999) Chemokines and activated macrophages in HIV gp120-induced neuronal apoptosis. Proc Natl Acad Sci U S A 96:8212–8216

18. Eugenin EA, D'Aversa TG, Lopez L et al (2003) MCP-1 (CCL2) protects human neurons and astrocytes from NMDA or HIV-tat-induced apoptosis. J Neurochem 85:1299–1311

19. Ghezzi S, Noolan DM, Aluigi MG et al (2000) Inhibition of CXCR-3-dependent HIV-1 infection by extracellular HIV-1 Tat. Biochem Biophys Res Commun 270:992–996

20. Magnuson DS, Knudsen BE, Geiger JD et al (1995) Human immunodeficiency virus type 1 tat activates non-N-methyl-o-aspartate excitatory amino receptors and causes neurotoxicity. Ann Neurol 37:373–380

21. Bonavia R, Bajetto A, Barbero S et al (2001) HIV-1 Tat causes apoptosis death and calcium homeostasis alterations in rat neurons. Biochem Biophys Res Commun 288:301–308

22. Haughey NJ, Cutler RG, Tamara A et al (2004) Perturbation of sphingolipid metabolism and ceramide production in HIV-dementia. Ann Neurol 5:257–267

23. Kruman LL, Nath A, Mattson MP (1998) HIV-1 protein Tat induces apoptosis of hippocampal neurons by a mechanism involving caspase activation, calcium overload, and oxidative stress. Exp Neurol 154:276–288

24. Nath A, Haughey NJ, Jones M et al (2000) Synergistic neurotoxicity by human immunodeficiency virus proteins tat and gp120: protection by memantine. Ann Neurol 47:186–194

25. Hurtrel M, Ganiere JP, Guelfi JF et al (1992) Comparison of early and late feline immunodeficiency virus encephalopathies. AIDS 6:399–406

26. Thormar H (2005) Maedi-Visna virus and its relationship to human deficiency virus. AIDS Rev 7:233–245

27. Lackner AA, Veazey RS (2007) Current concepts in AIDS pathogenesis: insights from the SIV/macaque model. Annu Rev Med 58:461–476

28. Toggas SM, Masliah E, Rockenstein EM et al (1994) Central nervous system damage produced by expression of the HIV-1 coat protein gp120 in transgenic mice. Nature 367:188–193

29. Bruce-Keller AJ, Turchan-Cholewo J, Smart EJ et al (2008) Morphine causes rapid increases in glial activation and neuronal injury in the striatum of inducible HIV-1 Tat transgenic mice. Glia 56:1414–1427

30. Agrawal L, Louboutin JP, Reyes BAS et al (2006) Antioxidant enzyme gene delivery to protect from HIV-1 gp120-induced neuronal apoptosis. Gene Ther 13:1645–1656

31. Louboutin JP, Reyes BAS, Agrawal L et al (2007) Strategies for CNS-directed gene delivery: in vivo gene transfer to the brain using SV40-derived vectors. Gene Ther 14:939–949

32. Louboutin JP, Agrawal L, Reyes BAS et al (2007) Protecting neurons from HIV-1 gp120-induced oxidant stress using both localized intracerebral and generalized intraventricular administration of antioxidant enzymes delivered by SV40-derived vectors. Gene Ther 14:1650–1661

33. Louboutin JP, Agrawal L, Reyes BAS et al (2012) Gene delivery of antioxidant enzymes inhibits HIV-1 gp120-induced expression of caspases. Neuroscience 214:68–77

34. Nosheny RL, Bachis A, Acquas E et al (2004) Human immunodeficiency virus type 1 glycoprotein gp120 reduces the levels of brain-derived neurotrophic factor in vivo: potential implication for neuronal cell death. Eur J Neurosci 20:2857–2864

35. Louboutin JP, Agrawal L, Reyes BAS et al (2009) HIV-1 gp120 neurotoxicity proximally and at a distance from the point of exposure: protection by rSV40 delivery of antioxidant enzymes. Neurobiol Dis 34:462–476

36. Louboutin JP, Agrawal L, Reyes BAS et al (2010) HIV-1 gp120-induced injury to the blood-brain barrier: role of metalloproteinases 2 and 9 and relationship to oxidative stress. J Neuropathol Exp Neurol 69:801–816

37. Louboutin JP, Reyes BAS, Agrawal L et al (2010) Blood-brain barrier abnormalities caused by exposure to HIV-1 gp120 - protection by gene delivery of antioxidant enzymes. Neurobiol Dis 38:313–325

38. Louboutin JP, Reyes BAS, Agrawal L et al (2010) HIV-1 gp120-induced neuroinflammation: relationship to neuron loss and protection by rSV40-delivered antioxidant enzymes. Exp Neurol 221:231–245

39. Agrawal L, Louboutin JP, Marusich E et al (2010) Dopaminergic neurotoxicity of HIV-1 gp120: reactive oxygen species as signaling intermediates. Brain Res 1306:116–130

40. Louboutin JP, Reyes BAS, Agrawal L et al (2011) HIV-1 gp120 upregulates matrix metalloproteinases and their inhibitors in a rat model of HIV encephalopathy. Eur J Neurosci 34:2015–2020

41. Louboutin JP, Agrawal L, Reyes BAS et al (2009) A rat model of human immunodeficiency virus-1 encephalopathy using envelope glycoprotein gp120 expression delivered by SV40 vectors. J Neuropathol Exp Neurol 68:456–473

42. Louboutin JP, Agrawal L, Reyes BAS et al (2014) Oxidative stress is associated with neuroinflammation in animal models of HIV-1 Tat neurotoxicity. Antioxidants 3:414–438

43. Agrawal L, Louboutin JP, Reyes BAS et al (2012) HIV-1 Tat neurotoxicity: a model of acute and chronic exposure, and neuroprotection by gene delivery of antioxidant enzymes. Neurobiol Dis 45:657–670

44. Ikonomidou C, Bosch F, Miksa M et al (1999) Blockade of NMDA receptors and apoptotic neurodegeneration in the developing brain. Science 283:70–74

45. Ribe EM, Serrano-Saiz E, Akpan N et al (2008) Mechanisms of neuronal death in disease: defining the models and the players. Biochem J 415:165–182

46. Madden SD, Cotter TG (2008) Cell death in brain development and degeneration: control of caspase expression may be key! Mol Neurobiol 37:1–6

47. Sims NR, Muyderman H (2010) Mitochondria, oxidative metabolism and cell death in stroke. Biochim Biophys Acta 1802:80–91

48. Broughton BRS, Reutens DC, Sobey CG (2009) Apoptotic mechanisms after cerebral ischemia. Stroke 40:e331–e339

49. Rohn TT (2010) The role of caspases in Alzheimer's disease: potential novel therapeutic opportunities. Apoptosis 15:1403–1409

50. Petito CK, Roberts B (1995) Evidence of apoptotic cell death in HIV encephalitis. Am J Pathol 146:1121–1130

51. Fujikawa DG (2015) The role of excitotoxic programmed necrosis in acute brain injury. Comput Struct Biotechnol J 13:212–221

52. Bottone MG, Fanizzi FP, Bernocchi G (2015) In vivo and in vitro immunohistochemical visualization of neural cell apoptosis and autophagy. In: Merighi A, Lossi L (eds) Immunocytochemistry and related techniques, vol 101, Neuromethods. Springer Protocols. Springer Science + Business Media, Humana Press, New York, NY, pp 153–178

53. Agrawal L, Louboutin JP, Strayer DS (2007) Preventing HIV-1 Tat-induced neuronal apoptosis using antioxidant enzymes: mechanistic and therapeutic implications. Virology 363:462–472

54. Louboutin JP, Marusich E, Fisher-Perkins J et al (2011) Gene transfer to the Rhesus monkey brain using SV40-derived vectors is durable and safe. Gene Ther 18:682–691

55. Louboutin JP, Chekmasova AA, Marusich E et al (2010) Efficient CNS gene delivery by intravenous injection. Nat Methods 7:905–907

56. Louboutin JP, Reyes BAS, Agrawal L et al (2012) Intracisternal rSV40 administration provides effective pan-CNS transgene expression. Gene Ther 19:114–118

57. Paxinos G, Watson C (1986) The rat brain in stereotaxic coordinates, 2nd edn. Academic, New York, NY

58. Louboutin JP (2015) Immunocytochemical assessment of blood-brain barrier structure, function, and damage. In: Merighi A, Lossi L (eds) Immunocytochemistry and related techniques, vol 101, Neuromethods. Springer Protocols. Springer Science + Business Media, Humana Press, New York, NY, pp 225–253

Neuromethods (2016) 115: 245–246
DOI 10.1007/978-1-4939-3640-3
© Springer Science+Business Media New York 2016

INDEX

Printed in the United States
By Bookmasters